Warrior Ways

Warrior Ways

Explorations in Modern Military Folklore

Edited by
Eric A. Eliason and Tad Tuleja

UTAH STATE UNIVERSITY PRESS

❖

LOGAN

© 2012 by the University Press of Colorado
Published by Utah State University Press
An imprint of University Press of Colorado
5589 Arapahoe Avenue, Suite 206C
Boulder, Colorado 80303

The University Press of Colorado is a proud member of

 The Association of American University Presses.

The University Press of Colorado is a cooperative publishing enterprise supported, in part, by Adams State University, Colorado State University, Fort Lewis College, Metropolitan State University of Denver, Regis University, University of Colorado, University of Northern Colorado, Utah State University, and Western State Colorado University.

ISBN: 978-0-87421-903-6 (paper)
ISBN: 978-0-87421-904-3 (e-book)

Library of Congress Cataloging-in-Publication Data

Warrior ways : explorations in modern military folklore / edited by Eric A. Eliason and Tad Tuleja.
    p. cm.
  Includes bibliographical references and index.
  ISBN 978-0-87421-903-6 (pbk.) — ISBN 978-0-87421-904-3 (e-book)
1. Soldiers—Folklore. 2. Armed Forces—Folklore. 3. Military art and science—Folklore. I. Tuleja, Tad, 1944- II. Eliason, Eric A. (Eric Alden), 1967-
  GR517.W37 2012
  398.276—dc23
                                          2012033615

For our fathers

COLONEL C. DANIEL ELIASON
United States Air Force, Retired
(Vietnam 1970)

CAPTAIN THADDEUS V. TULEJA
(1917–2001)
United States Navy
(World War II)

Contents

PART III: BELONGING

PART IV: REMEMBERING

Acknowledgments

Preliminary versions of several essays in this volume appeared in a double panel on military folklore presented at the 2010 American Folklore Society meeting in Nashville. We thank Lydia Fish for suggesting the panel and Margaret Mills for serving as a discussant. John Alley, former acquiring editor at Utah State University Press, invited our submission of the manuscript, and the press's director, Michael Spooner, saw it into production with his customary acumen. For their insightful feedback, we are also grateful to Simon Bronner and James Leary and to two anonymous peer reviewers.

Warrior Ways

Introduction

Modern Military Folklore
Retrospect and Prospects

Tad Tuleja and Eric A. Eliason

As countless recruits have been told by their boot camp instructors, "There's a right way, a wrong way, and the Army way. You will learn to do it the Army way." The familiar adage reflects not only the armed forces' perhaps exaggerated reputation for inflexibility, but the less debatable fact that the members of military organizations, like the members of other occupational folk groups, are enculturated into characteristic behavior patterns that identify them as group members, create solidarity, help them manage stress, and distinguish them from those not in the group. These patterns, which are observable in everything from dress to discourse, from jokes and songs and stories to institutional memory, comprise the capacious "it" that members perform the Army way or, if they are in another service branch, according to the customs and traditions of that branch. Army, Navy, Air Force, Coast Guard, Marines: each has its own occupational folklore—and there are often specialized variants of that lore at unit levels, such as among Army nurses or Navy Seals.

Examples of military folklore are scattered throughout the vast literature on war and warriors, going back to Homer and the Old Norse sagas; indeed, it would be difficult to write about any military activity without reference to warriors' vernacular speech, dress, daily routines, customs, and traditions. But as a subject inviting academic exploration in its own right, this particular form of occupational folklore, at least in English, has entered the scholarly purview only fairly recently. Tristram Coffin's analytical index[1]

1. Tristram P. Coffin, *An Analytical Index to the Journal of American Folklore* (Philadelphia: American Folklore Society, 1958).

DOI: 10.7330/9780874219043.c00   1

(1958) to the first seventy years of the *Journal of American Folklore* contained occasional entries on "heroes" and "pirates," but none on "soldiers" or "military folklore"; Joan Perkal's similar index[2] (1966) to the first quarter century of *Western Folklore* mentioned only four entries in these latter categories. It would take World War I to turn folklorists' attention to this neglected arena, World War II to bring that attention to critical mass, and Vietnam to bring it to fruition. *Warrior Ways*, building on this still maturing arena, is an attempt simultaneously to consolidate existing insights and to push the boundaries of the field into unexplored territory.

Our focus is intentionally narrow. Volumes could no doubt be written on the folklore of Roman legionnaires, Turkish janissaries, Lakota dog soldiers, or British infantrymen under the Raj. We leave such worthy tasks to more cosmopolitan souls. As North American scholars, we focus modestly and decisively on the modern period (that is, the twentieth and twenty-first centuries), on English-speaking (mostly American) warriors, and—even more decisively—on the lore produced by those warriors themselves, rather than on institutions mandating customs and traditions. In the essays assembled here, we hope to illuminate the ways in which members of the armed services creatively utilize various forms of expressive culture when they—as Jay Mechling put it nicely in another context—"come together to make meaning in their lives."

Perhaps predictably, it was a major conflict—World War I—that sparked scholars' initial interest in military folklore, and that led, even as the guns were still rattling, to some excellent early collections of modern warrior folkways. One of these, Fraser and Gibbons's *Soldier and Sailor Words and Phrases* (1968 [1925]), pioneered an attention to military slang that has commanded scholarly interest ever since. This volume remains a standard guide to British vernacular speech of the Great War, continuing to elicit respect even after the appearance of a later standard, *The Long Trail* (1965), by John Brophy and the dean of British slang scholars, Eric Partridge.

The Long Trail covered both slang and song, and in its attention to the latter it built on a rich tradition of occupational folksong collections—many of them by World War I veterans—that dated back to the conflict itself. From the British Expeditionary Force came Lieutenant F. T. Nettleingham's *Tommy's Tunes* (1917), a handbook of songs favored by "the boys at the front" and including numerous parodies of popular tunes.

---

2. Joan Ruman Perkal, compiler, "Index," *Western Folklore* 25, no. 4 (1966): 297–304.

From the American Expeditionary Force came *Ye AEF Hymnal* (1918), published anonymously in France with the doughboys' favorites. Shortly after the war, the *Journal of American Folklore*—still in its infancy—ran articles applying the then popular theory of communal composition (the "singing, dancing throng") to soldiers' songs, especially the best-known World War I example, the hydra-headed, ribald "Hinky-Dinky Parlez-Vous" (Hench 1921; see also Pound 1923; Cary 1935). This same oeuvre—rich in parodic license and good-natured grumbling at the "Army way"—inspired several good later collections: *The Songs My Mother Never Taught Me* (Niles, Moore, and Wallgreen 1929), Edward Dolph's *Sound Off!* (1929), Dorothea York's *Mud and Stars* (1931), and Melbert Cary's definitive catalog of "Hinky-Dinky" verses, *Mademoiselle from Armentieres* (1930, 1935). Other valuable sources of World War I songs are Niles, Moore, and Wallgreen's *Singing Soldiers* (1927), which focused on African American "doughboys," and Max Arthur's short, pungent volume, *When This Bloody War Is Over* (2001).

World War II generated its own complement of song compilations, notably an update of Dolph (1942; a revised and expanded edition of the original compilation published in 1929), classic collections by Hamish Hamilton (1945) and Eric Posselt (1943); and, from Great Britain, retrospective volumes by Martin Page (1973, 1976) and C. H. Ward-Jackson (1945). That conflict also widened the scope of those interested in military folklore. The first year of the war, for example, saw two notable instances of the "regimental potpourri" genre: Cyril Field's *Old Times under Arms: A Military Garner* (1939), and a third edition of Leland Lovette's Naval Institute standard *Naval Traditions and Usage* (1939). These were followed by similar honorific compendia on the customs of the British (Edwards 1950) and American (Boatner 1976) armed forces.

An early exception to this "top-down" perspective was Agnes Underwood's 1947 article "Folklore from G.I. Joe," in the then infant *New York Folklore Quarterly*. Underwood was teaching composition at Russell Sage College to young veterans who were among the first beneficiaries of the new GI Bill. Among the items of World War II folklore they shared with her, with the obscenities chivalrously omitted, were examples of slang, songs, jokes, cadences, slogans, superstitions, and prayers—plus a variant of the popular shaggy-dog story, "The Kluge Maker," which was also recorded by Jansen (1948) and Dorson (1977 [1959]). A fellow military folklorist, Les Cleveland, later (1987) indexed Underwood's collection and rightly lauded her as a pioneer in the "emic" approach. Her article, and his response, helped to establish the New York journal as a small but distinguished venue for military folklore articles (see also Roulier 1948; Keith 1950; Koch 1953).

The 1940s also saw heightened scholarly interest in military language, of both the bureaucratic variety that Elinor Levy, in this volume, calls *officialese*, and the foot soldiers' slang that she dubs *enlistic*. Descriptive compilations in this period included Elbridge Colby's *Army Talk* (1942), John Riordan's "American Naval 'Slanguage' in the Pacific" (1946), and Joseph Roulier's "Service Lore: Army Vocabulary" (1948), which argued that linguistic ingenuity helped servicemen adapt to institutional rigidity. The postwar period—with pop Freudianism on the rise—also saw the first glimmerings of psychological analysis, with sociologist Frederick Elkin's argument (1946) that military lingo expressed soldiers' conflicted adjustment to routine and authority. Later psychological analyses included Thorpe's (1967) sensitive study of his fellow flyers' mindset, an examination of the "saltpeter" myth by Rich and Jacobs (1973), and Kenagy's (1978) study of pilots' sexualized speech.

Like previous conflicts, the war in Vietnam both reinforced existing military speech and elicited creative additions to its repertoire. An early example of interest in this phenomenon was Joseph Tuso's reproduction in *Folklore Forum* of the mock interview "What the Captain Means." Reprinted by, among others, Burke (2003, 2004), Pratt (1984), and Tuso (1990) himself, it remains a classic illustration of institutional obfuscation. Subsequent studies of Vietnam-era jargon have included Cornell's (1981) "G.I. Slang in Vietnam" and helpful dictionaries by Clark (1990) and Reinberg (1991). Dictionaries with a wider historical focus have been produced by Elting, Cragg, and Deal (1984) and Dickson (2004). The latter work brings the collection of military slang into the twenty-first century, as does Bay's (2007) guide to "milspeak" and Robson's (2008) well-annotated compendium of naval slang.

The war in Vietnam also dramatically expanded the production of military folksongs and of studies devoted to their analysis. In the Vietnam era especially, it is important to distinguish between songs written about the military by civilians and songs made by military personnel themselves. The former are ably documented in Serge Denisoff's *Songs of Protest, War, and Peace* (1973), a standard source for music of the antiwar movement. For the latter, the indispensable authorities are Les Cleveland and Lydia Fish. Cleveland was a New Zealand airman who wrote about the occupational folksongs of soldiers serving in conflicts from World War II to Vietnam. His essay on the "folklore of the powerless" (1985) and his book *Dark Laughter* (1994a) showed how soldiers used such songs to protest against authority and manage fear. His essays on Vietnam's "occupational folk tradition" (2003 [1988]) and "singing warriors" (1994b) compared US and New Zealand reactions to the Vietnam engagement. All of his work is richly contextualized, well annotated, and insightful.

Lydia Fish, a Buffalo State College professor who worked with Cleveland on early—and still useful—bibliographies of military lore, continued her colleague's interest in the Vietnam era. Through her management of the Vietnam Veterans Oral History and Folklore Project (VVOHFP), she has become an unrivaled expert on the occupational folksongs of that era. The project's website is an essential first stop for anyone researching Vietnam-era folklore. In addition, the "Articles and Papers" link on Fish's website (http://faculty.buffalostate.edu/fishlm) includes her works "Songs of Americans" (1993) and "Songs of the Air Force" (1996) during the war. Fish has written an appreciative assessment (1989) of Air Force officer Edward Lansdale (1908–1987), who during the war sent valuable collections of such songs to the Library of Congress. As VVOHFP director, she also produced a seminal recording of 1960s military folksong, the Flying Fish CD *In Country* (1991).

The Vietnam War holds an enduring fascination for scholars of the baby boom generation, and this has elicited, in addition to Fish's work, two online resources of special note. In 1988, humanities scholar Kali Tal founded the electronic journal *Viet Nam Generation* and shortly thereafter *The Sixties Project*, a website hosted by the University of Virginia that includes essays, poetry, and personal narratives about that turbulent decade. Of particular interest is the journal's special issue "Nobody Gets off the Bus: The Viet Nam Generation Big Book," also available in book form (see Baky 1994). The second resource is the journal *New Directions in Folklore*, published at Temple University. Its special issue on military folklore, guest edited by Carol Burke (2003), includes useful essays on occupational folksong by Fish, Cleveland, and Martin Heuer as well as essays on military speech by Burke, gun lore by Richard Burns, and World War I poetry by Timothy Rives.

While the scholarship on military folksongs such as off-hours ballads and parodies is copious, the literature on the most occupationally integrated American military folk music, marching and running cadences, or "Jody calls," has by contrast been limited. Like Jodies themselves, this scholarship has been provocative. Work in this area began in 1965, when former para-trooper George Carey, then teaching at Middlebury College, published a collection of marching chants he had heard from members of the 101st Airborne Division. Two years later, Bruce Jackson, asking "What Happened to Jody?" explored the roots of this para-musical genre in African American culture (1967). In the 1980s, Sandee Shaffer Johnson published two collections of expurgated texts at the ironically named Daring Press (1983, 1986). Subsequent studies by Burke (1989), Knight (1990), and Trnka (1995) emphasized the chants' aggressive and misogynistic implications. More recently, Richard Allen Burns has investigated cadences as part of "gunlore" (2003) and, in a volume devoted to "recovered, represented, and reimagined" folksongs, has stressed the role of drill instructors as military tradition bearers struggling to adapt the lubricious material to changing circumstances—including the entry of women into the armed forces (2006). His chapter in this volume builds on these earlier explorations.

The material culture of modern warriors—the stuff they made and, in Tim O'Brien's famous phrase, the "things they carried"—has received remarkably little attention from English-speaking folklorists. Information on collectible items is available in antiques magazines (see, for example, Smyth 1988), but scholars have yet to turn much attention to such items of soldierly interest as graffiti, good luck charms, mementoes, pinups, barracks décor, and creative dress style. Notable exceptions include Chittenden's (1989) study of a Vietnam veteran's vernacular art, the recollections by Dewhurst (1988) and Sossaman (1989) of so-called Pleiku jackets, and Bernadene Ryan's (2011) examination of challenge coins. The observations in this volume by Gilman on military dress, Mechling on soldiers' photography, Eliason on their appropriation of Afghan decoration, and Burke on battlefield talismans make forays into this largely unexplored territory. We hope that they may inspire future investigations.

Material culture is only one of several areas in which modern warrior ways invite further attention. One area with obvious research potential is the lore of men and women engaged in conflicts that postdate what seems to be, for folklorists no less than historians, the perennially fascinating terrain of Vietnam. Since the fall of Saigon in 1975, the United States has been almost uninterruptedly engaged in missions abroad, yet none of them has elicited much attention from folklorists. The *Journal of American Folklore*, for example, has published only one article on folk (as distinct from media) expressions generated by the Second Iraq War, a study of a father's memorial to his fallen son (Pershing and Bellinger 2010). The essays in this volume by Gilman, Oswald, Burke, and Eliason focus attention on the veterans of Iraq and Afghanistan, but there is much more to be done in this area, as there is in the study of America's limited interventions in Lebanon, Grenada, Panama, Haiti, Bosnia, Kosovo, and Somalia.

There is more to be done especially on the conflict in Korea. As the sixtieth anniversary of that prelude to Vietnam approaches, it remains, for folklorists no less than the general public, America's "forgotten war." Dan Cragg's (1980) study of sarcastic procedural signs, or "prosigns," draws on his experience in Korea, and George Carey's seminal essay (1965) on marching chants focuses on those current in the Korean War era. But to our knowledge no studies have yet appeared in folklore literature of such potentially rich subjects as war brides, brainwashing, Douglas MacArthur anecdotes, or the adoption by US (and UN) forces of Korean martial arts.

Students of military folklore might say more about race, class, and their interaction. While the experiences of American warriors of color such as the Tuskegee Airmen, Navajo code breakers, and the Vietnam War's "bloods" have garnered some popular attention, they have not been adequately scrutinized by folklorists. As suggested by Tom Vennum and Mickey Hart's CD of Ho Chunk (Winnebago) military songs (1997), the lore of Native Americans in the US services provides an especially fertile field for further research. Folklorists might also help to analyze the hints of social tension—evidence of a muted class warfare—that appear in songs, tales, and jokes about the "brass" from at least World War I onward.

We might also pay more attention to the changing gender composition of the military community. Carol Burke has set a high standard in this regard, and feminist scholars have done creditable sociological work on the gendering of the military. But much more could be done with regard to women warriors' own folk culture—on the ways in which their experiences are reflected in story, song, dress, and other expressions. The same point could be made about the experiences of gay and lesbian members of the armed forces. Mickey Weems's and Eric Eliason's essays in this volume break interesting ground by addressing this long marginalized (and suppressed) segment of the warrior class; in the post–Don't Ask, Don't Tell era, there is an opening for folklorists to follow their lead.

Another understudied area is the naval services. The formal "ways" of the seagoing services have long been amply documented in official sources such as Lovette's *Naval Traditions and Usage* (1939). But, with the exception of the slang studies noted above (Riordan 1946; Robson 2008), Simon Bronner's study of equator-crossing rituals (2006), and Burke's work on Naval Academy traditions (2004), scholars have paid relatively little attention to vernacular Navy or Marine traditions, and even less to those of the United States Coast Guard. Gillespie's chapter in this volume takes a step toward correcting that oversight. We trust it may spark renewed interest in "sea service" folkways.

More, too, might be done to explore the folkways of those in auxiliary or support roles for the military proper. Chief among these would be military spouses—a group whose culture Young explores in this volume. Other "paramilitary" groups deserving of study might include USO performers and staff, organizations of Gold Star mothers, military contractors, and students in military academies. The folkways of these corollary groups, worthy of study no doubt in their own right, might also enrich our understanding of warriors' folkways.

Another area for research is the role of the new media made possible by the Internet in transmitting, manipulating, and debating military folklore. Today, "warrior ways" are made public knowledge, and thus available for discussion, through a network of government websites, other "official" websites (such as those managed by private "patriotic" organizations), Facebook and other social networking pages, YouTube postings, and forums and chat rooms numbering in the hundreds if not thousands. Information and exchanges posted in these electronic venues provide rich insight into public images of the military, and into the contentious, often vitriolic debates about the activities in which they are engaged. We hope that the brief invocation of these venues by Gillespie, Weems, and Tuleja in this volume may point to their potential value as arenas of further research.

Finally, a word on humor. Scarcely a chapter in this book fails to indicate, directly or by implication, the centrality of humor to military life, and its fruitfulness as an index of warriors' worldviews. Dictionaries of slang, song collections, cadences, stories, material culture: anywhere one finds military folk expression, a sense of sardonic resignation, often ribald, is not far afield. Examples have appeared in volumes such as those listed in our brief "Humor" bibliography as well as in venues such as *Reader's Digest*'s perennially popular "Humor in Uniform" feature—now appearing under the feature title "Off Base." Little critical analysis, however, has been brought to bear on military humor, and this is a field in which psychosocial approaches could be especially valuable. Surely a close reading of typical soldiers' jokes, puns, and humorous stories could be as useful in illuminating warriors' worldviews as are existing studies of their vernacular speech and traditions.

With Western militaries pushing the bleeding edge (literally) of technological change, military folkloristics must evolve with the changing ways of war. In line with this imperative, *Warrior Ways* continues the tradition of military folklore scholarship recounted above but also moves the field forward, looking at issues and addressing questions raised by changing conditions. Some of its contributors are senior scholars; others have their first publication in this volume. Some employ speculative and theoretical perspectives; others are descriptive and historical. They are women and men who display a wide variety of relationships to the military. Several have service backgrounds, some in combat abroad. Others are parents, offspring, and/or in-laws of military personnel. Some have worked for the military without being in the service themselves. All have had close contact with military people—in fieldwork, as teachers, or some other form of involvement.

The contributors also represent a variety of political, religious, and moral views about military service in general, and they hold as many different opinions about current and past conflicts as there are conflicts and contributors. However, while some of these views may manifest in an ancillary manner in some chapters, politicized scholarship is not what this collection is about. Each contributor seeks to understand and present some aspect of military expressive culture and analyze it critically but sympathetically while remaining cognizant of the sacrifices made by soldiers, sailors, airmen, and Marines.

At times, some of this book's contributors may present seemingly contradictory observations and conclusions. We don't try to reconcile them. Individually and collectively, we try to avoid the logical error of assuming that any partial glimpse into the varieties of military experience—like the proverbial blind men's grasping at parts of the elephant—can ever capture the complexity of the whole.

This expansive and, we hope, nonpolarizing approach is not, of course, universal. Despite the natural limits of observational and interpretive capacity and the uncircumscribable complexity of military experience, societies still tend to maintain rigid master narratives about their soldiers' experiences. In the past, for example, war was widely seen as a test of bravery and an opportunity for glory, and military service was seen as a duty one stoically undertakes on principle or for love of country. A rival narrative—currently fashionable although generally unpopular among veterans—is of soldiers as idealistic innocents whom we owe sympathy and support since they have been victimized in our name by political leaders who have sent them on psychologically damaging errands. Neither of these narratives captures the nuances of lived experiences as revealed in *Warrior Ways*.

Nor are those nuances captured by what are probably the two most durable of military stereotypes. One is the notion of the warrior as an amoral automaton, either made an unthinking killer through dehumanizing training or seeking a way to legitimately channel innate and perverse violent tendencies. In truth, because of their extraordinary responsibilities and the often harrowing nature of their work, soldiers may well deeply experience richer and more tormenting emotional and moral dilemmas than their civilian counterparts. Also, in Western militaries, where discipline and human rights are core values, those with undisciplined violent tendencies tend to get weeded out during recruitment or training. That there are sometimes well-publicized lapses from these core values does not invalidate this general observation.

The second common stereotype—that militaries draw disproportionately from the economically disadvantaged—assumes that service is an unjust burden rather than a sought-out honor. This stereotype is also untrue. In the United States military, 25 percent of all recruits come from the wealthiest 20 percent of eighteen- to twenty-four-year-olds, while only 11 percent come from the poorest 20 percent.[3] Recruits sign up for many reasons: to pay for school, as a family tradition, out of patriotism, for the adventure, for the training, and for the sense of being part of something bigger than themselves. This book testifies to that variety of motivation and suggests that military people's worldview—that central element of every folk group's experience— is similarly complicated.

The multiple truths of soldiers' experiences lie between and beyond pat master narratives, and are as varied and complex as are individual soldiers, their units, their missions, and the areas of operation in their respective conflicts. But master narratives are powerful, and soldiers whose experiences do not easily fit them often learn to adjust their own stories or to keep quiet, because deviation from a popular narrative can lead the unreflective to assume that the soldier himself or herself is deviant. This presents a challenge to scholars attempting to understand how military lives are lived from the

---

3. Ann Marlowe, "The Truth about Who Fights for Us," *The Wall Street Journal*, September 27, 2011.

inside, rather than in conventional media representations. Our contributors have attempted to let soldiers speak for themselves and invite readers to compare their actual expressions to cultural expectations

One of the ways social scientists seek to confound stereotypes and master narratives is by turning to native, or reflexive, ethnographers. This volume's first three chapters are all authored by folklorists whose service with the military goes beyond participant observation. One is a civilian Air Force employee. Two deployed to Afghanistan—one as a civilian contractor and one as a national guardsman. Each focuses on the most dramatic of military situations: in-country wartime deployment or, to use the term common among US soldiers in the Middle East, "playing in the sandbox."

Even with the multiple deployments of our current wars, soldiers don't spend most of their time downrange. And even when deployed there is much truth to the old saying "War is 90 percent boredom and 10 percent terror." But it is combat roles that most distinguish soldiers from civilians, and the unique circumstances of deployment produce distinctive expressive culture. In her essay on "the things they carry," Carol Burke explores the role of personal talismans as sources of magical comfort to soldiers fighting in Afghanistan. Based on her year's deployment, her essay shows how in an environment where the greatest threat may come from unconventional weapons such as IEDs, soldiers carry a variety of good luck charms—some brought from home, others obtained in country—intended to increase their chances of survival. The psychological value of these protective objects, she shows, sometimes has less to do with their owners' belief in their inherent potency than with the fact that they have been acquired as gifts, often from more seasoned soldiers. Their true magic for soldiers is how they demonstrate mutual caring among warriors.

Justin Oswald begins by examining what is arguably the most common of all military folk genres, the personal narrative. He reveals the role of the much-feared "camel spider" in stories that US soldiers tell about their Middle East deployments. These fearsome animals (they are not technically spiders, but "solifuges") are notorious for their speed, stealth, painful bite, and uncanny ability to penetrate the perimeters of tents and sleeping bags. Oswald shows how these characteristics make them a suitable metaphor for the indigenous enemy, and how staged "death matches" between them and scorpions provide a theatrical catharsis for those in harm's way. Camel spider stories, which often reference official warnings about the creatures, also serve as initiatory bonding vehicles for new recruits about to enter a danger zone.

Eric Eliason invites us to consider a military folk group not as a passive object of study but as a dynamic player in the creation and analysis of expressive culture. Drawing on his experience as a chaplain serving with US Special Forces in Afghanistan, he shows how these military professionals and the local Pashtun people have cross-fertilized each other's folkways as they cooperate in reconstructing and providing security for the region. His essay explores the ways in which Afghan traditional culture has influenced the practices of American soldiers and how they have developed a descriptive and analytical discourse to understand it. He focuses particularly on the sense-making American soldiers have done regarding Afghan folk theology, vernacular architecture, vehicle and tool ornamentation (what soldiers call "jinglefication"), and traditions of male bonding that are sometimes homoerotic, such as "Man-Love Thursday." In shifting the analytical perspective from that of scholars to that of the "folk" being studied, Eliason proposes the idea of "folk-folkloristics" as the bottom-up practice of folklore analysis done by folk groups other than folklorists.

The book's second section, which focuses on various modes of "sounding off," begins with a chapter on the most effective means of developing unit cohesion and group identity that the military has ever employed: the cadence calls that were first popularized in the military by Sergeant Willie Duckworth in World War II. Folklorist and retired Marine Rick Burns examines cadence singing's most common character, a certain "Jody" who is sweet-talking your girl and driving your car while you are away at war. This figure's historical importance has loomed so large that cadences are themselves often called "Jodies." Burns explains that Jody's popularity stems not only from the fact that he articulates soldiers' common fear of being cuckolded (when this volume's editor Eric Eliason served in Afghanistan as a chaplain, twenty soldiers came to him with wife and girlfriend worries for every one that came for combat stress-related issues), but also because Jody's attack on fidelity mirrors soldiers' own sexual waywardness. The same soldier who curses Jody one day may on another occasion spout the common saying "What happens TDY stays TDY" (TDY is a "temporary duty assignment" away from home base). This saying articulates a cavalier attitude toward sexual fidelity that, Burns suggests, may elicit guilt that cadences transfer onto Jody. This essay also examines the prevalence of death as a cadence call motif and underscores one of the most powerful insights of folklore as a discipline—that close reading of the texts the folk produce themselves can lead to greater ethnographic understanding than the theories and speculations that folklorists bring to the field.

Elinor Levy shifts from folksong to folk speech. Soldiers' lingo, like any occupational argot or subcultural cant, can be difficult for outsiders to penetrate. "Military speak," with its profusion of acronyms, slang, and self-referential allusions, has often been accused of being purposefully obscure. While Levy suspects that obscurantism and self-serving euphemism play a role, she also demonstrates how her twin vocabularies of "officialese" and "enlistic" allow for self-expression and efficient communication. Officialese flows down from the Pentagon at the top, while enlistic bubbles up from the barracks and the field. Levy shows that competent performers must be fluent in both varieties, or registers, of military speech, and that those registers often overlap. For example, while officialese encourages the use of acronyms for streamlining communication, it probably did not anticipate the enlistic "FUBAR" and "SNAFU" descriptors that soldiers often use to describe the chaos and fog of war.

Angus Gillespie moves from Levy's general observations to the particular lingo of the Navy and Coast Guard, giving us a humorous look at a wide variety of terms and their functions. He shows how "sea service slang" reveals the love-hate relationship that servicemen and women have with the military. He also gives examples within the Coast Guard, by far the smallest of the services, of cultural and linguistic divisions between those who serve on white-hulled law enforcement cutters, red-hulled icebreakers doing scientific research, and black-hulled service and buoy repair vessels. Gillespie's findings underscore the fact that while there may be overarching pan-military cultural folklore and traditions, much military lore is specific to individual branches, occupations, or units. We should speak therefore not of military culture but of military cultures as we try to understand how multiple warrior ways (not the warrior way) function in different social moments and different locales.

The book's third section explores the complexities of who may, and who may not, be considered a member of the military community. An opening chapter by Mickey Weems pulls threads from gay rights history and his own experience as a Marine into a complex tapestry showing how the tense, oppositional, but often surprisingly cooperative and interrelated relationship between the US military and homosexual expressions works itself out in various customs, traditions, narratives, and folk sayings. In Weems's portrait, military culture appears as less simplistically homophobic than is often supposed. He shows that the ramrod-straight Marine standing next to you in formation might not be so straight when she goes home to her spouse.

In many ways, spouses are as much a part of military culture and community as those who serve; they have rightly been called "shadow warriors." The shared lot of frequent moves, long separations, and worries about safety can bring spouses together in informal friendship networks and military-sponsored "family readiness groups." Not without reason, spouses often gripe about challenges and express frustration with the inscrutable and often life-wrenching ways of the military. Yet they can also develop deep pride and appreciation, as evidenced by blue-star flags in front-door windows and yellow-ribbon bumper stickers on minivans. Kristi Young's chapter shows that the quintessentially military spirit of camaraderie emerges not just among soldiers but among spouses as well. Through carefully listening to her informants, she finds that traditional military and religious values can help spouses form supportive communities with an ethos of helpfulness and inclusion.

The effects of shared military experience and camaraderie can be so strong that even those who have come to actively oppose the wars they fought in often continue to relate to each other in military ways. Lisa Gilman has spent several years following a group of veterans involved with the antiwar "Coffee Strong" activist organization outside of Fort Lewis in Washington State. These young men are not the "longhaired hippie freaks" military people often stereotypically associate with Vietnam-era protests. Although they now oppose the wars in Iraq and Afghanistan, they continue to display military bearing, love getting together to shoot firearms, and still use military lingo in their conversations. While some might see it as ironic that the members of the Coffee Strong community sees their opposition to the war as emerging, at least in part, from the values of honor and respect for law that the military taught them, Gilman reveals the logic of their "oppositional positioning" as the foundation of their identity as a folk group.

The final three essays focus in various ways on how expressive culture can simultaneously anchor and stimulate warriors' reflections. In his study of "Colonel Bogey's March," Greg Kelley shows how an official military song became a worldwide folk tradition. Originally composed by a British Army band director in honor of an eccentric colonel, the song was a well-worn favorite in the early days of World War I. Reworked in parody by soldiers in many nations' militaries in several twentieth-century wars, it has escaped the military's grasp and become a bawdy children's folksong on playgrounds around the world. Its most famous version (from World War II) references the testicular abnormalities of the Nazi high command, beginning with the famous opening line, "Hitler has only got one ball."

Kelley demonstrates the ubiquity of this folksong parody in the numerous references to it in popular culture.

Jay Mechling also examines a military folkway's escape from military control—the democratization of photographic technology's impact on our understanding of soldiers' experiences. As making photographic images becomes cheaper, more ubiquitous, and harder to contain, how might our sense of what war is, and what soldiers' lives are like, change? Mechling suggests that one change is that soldiers' ambivalence about being in the service can now become better known despite the military's desire to manage information about it. Every soldier, not just the public affairs officer, has a camera for recording experiences. As Mechling's collection of soldier snapshots shows, they do so prolifically, leaving visual evidence not only of ambivalence but also of their most mundane, most typical, and most puzzling activities.

Like Greg Kelley, Tad Tuleja examines the history of a popular song and its folk variations. During the Vietnam War, popular music's references to the military were mostly negative. Barry Sadler's 1966 folksong-style hit "Ballad of the Green Berets" was the most notable counterexample. Sadler, a US Army Special Forces soldier himself, cowrote this song with journalist Robin Moore, and it became the theme music for *The Green Berets,* John Wayne's movie supporting US involvement in Vietnam. Like the Colonel Bogey march, it spawned many parodies and spin offs. Some were put to use stoking interservice rivalries while others were appropriated by the antiwar movement. Tuleja also shows how Sadler's "Ballad" remains a touchstone for debate (much of it on YouTube) about the Green Berets and about war, the military, and masculinity more generally.

These last three chapters in particular demonstrate how military folklore, often expressed in the vernacular creativity of soldiers, doesn't always remain in an ostensibly isolated subculture. Instead, like the folklore of any group, military lore can reveal to us our common humanity by delighting and disturbing, infuriating and inspiring, not only those deeply invested in, but also those peripherally touched by, warrior ways.

NOTE

All works cited in this introduction and not mentioned in footnotes appear in the Selected Bibliographies beginning on page 271.

Part I
Deploying

1

The Things They Bring to War

Carol Burke

Luck is always borrowed, never owned.

—Norwegian saying

Today's US Army soldiers deploy with three sets of grayish green ACUs (Army combat uniforms) or three sets of the more trendy "multicams"; with a helmet that comes with pads, strap, and cover; with three pairs of G.I. boots that soldiers mold to their feet by wearing them, when new, in the shower; with canteen and cover, trenching tool, "Gerber" (a utility knife), and gloves; with both goggles to protect from sandstorms and sunglasses ("eyepro") that are only marginally more stylish; with wet-weather gear, jackets and fleece for cold weather, and even subzero-weather gear and insulated underwear. The outerwear is of high quality, but soldiers complain that the basic uniform is far too heavy in the sweltering summer heat. They haul their Kevlar body armor (with attachments for neck, shoulders, and groin) and the hefty "small-arms protective inserts" (SAPI) plates that fit into the vest and protect against high-velocity rifle rounds, and two sleeping bags—one for warm weather, one for colder weather with the option of fitting the former inside the latter. Then there's the "bivy," the bivouac sack that insulates the sleeping bag from the cold and wet. Any soldier will tell you that the only way to sleep in Army sleeping bags is nude so that body heat will be reflected off the synthetic bag. They also carry an air mattress, a rain poncho, ammo pouches, a laundry bag, a waterproof washing bag, more protective gear, a "CamelBak" (water reservoir), and eating utensils. Everything, along with a few personal items, fits snugly into one rucksack,

DOI: 10.7330/9780874219043.c01   19

three hefty duffles, and an "assault pack," a small backpack for toiletries, laptop, underwear, socks, and towel for the several-day trip to their destination.

Among the personal items, soldiers bring books for the online courses they will complete during their year away and extra sheets to transform a bunk bed in a tent shared with eleven to twelve other soldiers into a private space. They bring pictures of lovers, husbands, wives, and children and metal bracelets with the names of fallen comrades from previous deployments. Because most forward-operating bases (FOBs) maintain generators to heat and cool every tent and plywood structure and to keep the lights on, the computers working, and the showers hot, the soldiers and civilians who live on them bring earplugs to drown out the constant drone of generators. Some need more help in getting to sleep, and they bring white-noise machines and doctor-prescribed medication to counteract the psychological scars from previous deployments.

Deployed soldiers sport their current unit's patch on their left shoulder and the patch from a previous deployment on their right shoulder. That's on the outside; on the inside they bring their freshly inked deployment tattoos or the designs of tattoos commemorating previous deployments. Marines typically have themselves inscribed with the Marine eagle, globe, and anchor or the motto "Death Before Dishonor" as "visible reminders of who they are," according to retired Marine Corps commander Colonel Mike Denning. Other soldiers wear images of skulls, flames, and weapons attesting to their power to vanquish whatever menace might come their way. Some soldiers who have come "downrange" (soldier-speak for deployment) to Iraq and Afghanistan have brought with them a second set of dog tags, what they refer to as their "meat tags," an exact image of their official dog tags inked onto their torsos. Since the single most deadly weapon in the insurgent's arsenal in these conflicts is the improvised explosive device, many soldiers make sure that their body, should it be blown up by an IED, can be distinguished from the remains of others. The stated purpose, however, is probably not the real function this practice serves. First, every squad leader and his platoon leader know who is going out on every mission and would instantly know who is missing. What's more, bodies would likely be charred and the "meat tag" unrecognizable were the explosion to leave only a torso. No, the ritual of going with buddies to get such a tattoo allows soldiers to acknowledge the worst that might happen, note that harsh fact on their flesh, and then get on with the mission. Meat tags function in the same way as macabre battlefield humor; they symbolically inoculate the soldier against thoughts that might otherwise incapacitate.

Those seasoned soldiers who have been this way before bring to war tales of other deployments. Sometimes they are simply short descriptions of a character they've encountered. One Marine, who asked that his name not be used, had served some hard time on previous deployments in both Iraq and Afghanistan and had lost close friends, but kept those stories to himself. He did, on the other hand, tell stories that showed his fondness for the locals he'd encountered. One of these colorful characters was affectionately named "Tooth" because he appeared to have only one tooth in his mouth, holding onto it as if it were a survivor's medal. Tooth, a groundskeeper at the US embassy in Kabul, loved all Marines, and those who worked at the embassy were fond of him, giving him one of their extra rank pins or a challenge coin. He sported them all. Whenever he spotted a Marine, Tooth would holler a jovial "Hey, Jarhead!" or "Hello, Devil Dog!" One day a Marine said, "Hey, Tooth, how old are you?" The man didn't know. "Were you ever a soldier?" the Marine asked. "Of course," said Tooth. "Did you fight the Russians?" said the Marine. Tooth dropped the lighthearted demeanor and looked seriously at him and said, "Yes." "Did you kill some Russkies?" asked the Marine. Tooth paused, then answered, "I've killed more Russians than cancer." Not only had Tooth fought the Russians; he had resisted the Taliban takeover as well. Because of his resistance, the Taliban came to his home and killed every member of his family in front of him. He was being tortured and would have been killed as well had a small unit of Marines not rescued him.

PERSONAL PROTECTION IN THE FACE OF DANGER

When civilians and soldiers go off to war they leave their families, their civilian friends, their favorite hangouts, and their possessions; they enter a walled community separated from the local population, which they regard as potentially hostile. It's a place where the unexpected is to be expected, where there is little one can do to mitigate risk other than rely on skills learned in training, stay alert, and keep a clean weapon. Even so prepared, there are no guarantees. War is about hidden danger waiting for the opportune time to present itself. In today's wars of counterinsurgency there is no territory to be taken, no mass of troops to eliminate. There are only the hearts and minds of a population to win. In areas where insurgents still maintain strength, the sympathies of this population are, of necessity, split. Despite their overwhelming military superiority, occupying forces in these

regions cannot protect the residents of every small village, so those villagers, who typically want little more than to work their land, to feed their
families, and to see their children grow and prosper, must live with what
they cannot change.

To move outside the protection of the base the American soldier wears
the best body armor, carries the best weapons, and rides around in the best
patrol vehicles. In spite of this, an insurgent can for a few bucks construct an
IED that will erase the advantage. In the face of such unpredictable danger,
those who deploy carry protective objects brought from home or acquired
while "in country." In this essay I will be discussing the use of such objects
in recent wars in Iraq and Afghanistan, but it is worth noting that taking
such tokens to war is not a recent development. Spanish soldiers in the late
1800s carried charms with "détente bala" (stop bullet) written on them.
World War II Japanese soldiers wore 1,000-stitch belts that sisters, mothers,
or wives had patiently assembled (figure 1.1).

According to tradition, every stitch of these belts had to come from a
different person, so Japanese women would stand at a busy market or inter-
section asking passersby to make a simple stitch. Most displayed no pattern
to the stitches; they simply fell in rows across the cloths, typically worn by
soldiers around their waists or heads. Some of these gifts came with a special
pocket in which could be inserted a good luck saying or a good luck coin—
a sort of amulet within an amulet. Other Japanese soldiers took with them
send-off banners emblazoned with the rising sun and inscribed by the family or friends who wished the soldier well. All were believed to bring protection to the wearer and ensure his return to his loved ones.

Talismans have also been used more recently in Thailand. On February
2, 2011, journalist Saksith Saiyasombut reported in the *Bangkok Post* (2011)
that Thawatchai Samutsakhon, Second Army chief in the Thai Army, feared
that the Cambodian troops stationed just across the border from his Thai
guards might be sending curses their way, so he decided to give his soldiers
extra protection. He issued all of them talismans with the image of a Thai
monk whom the Army chief revered, and instructed his Thai soldiers to
wear the protective devices on top of their body armor so that the monk's
image would face the enemy. The Cambodians on the other side would
not confirm the use of such unconventional weapons, but they would not
deny it either. Their spokesman said that the Cambodian soldiers would
likely perform rituals at a nearby temple in order to counteract the magic of
the Thai talismans. Although designed for border soldiers, such protective

1.1. A protective belt, as worn by Japanese soldiers in World War II.

devices are also showing up on the body armor of soldiers charged to face down antigovernment forces in Bangkok, the capital.

YOU DON'T HAVE TO BELIEVE FOR THE MAGIC TO WORK

Many soldiers, Marines, and officers bring to war the religious objects of their faith. As retired Marine colonel Mike Denning explained, "I wore a miraculous medal as a visible symbol of my faith, a touchstone" (2012). Although Denning obtained his medal on a trip to Lourdes with his wife, most amulets of war, including religious artifacts, are not purchased but are given by a friend or family member. Denning himself gave a young man about to deploy to Afghanistan a St. Michael medal. The archangel who tramples Satan and St. Christopher rank as the most popular saints among both Catholic and non-Catholic soldiers. The gift of an amulet circulates from one soldier to another, one officer to another, one Marine to another, as a token of the bond between those who have served and those who are currently serving.

Like Denning and other religious American military personnel, Afghan soldiers also carry with them objects that represent their faith. They wear necklaces inscribed with a verse from the Quran and necklaces with a gold or silver cylinder that holds a small scroll with a Quranic verse. They typically receive these tokens of their faith from a mother or grandmother. The wearing of these necklaces, like praying to saints, was outlawed during the reign of the Taliban, and the practice still invites controversies in some Islamic communities. Along with any religious artifacts that Afghan soldiers bring with them, they experience daily reminders of their faith. Although exempt from reciting their required prayers when out on a mission, Afghan soldiers on base typically stop what they are doing, pull down their prayer rugs, and recite their prayers in answer to the call to prayer that broadcasts from the base loudspeaker five times a day.

Their speech, too, is infused with references to Allah. The most common Afghan expression, a saying that at one time or another irritates most American commanders, is "Insha'Allah" (literally, "God willing"). To an Afghan, such a closing to an agreement to do something in the future attests to the overriding power of Allah, in whose hands everything rests, and the saying rhetorically gives the speaker cover in case he fails to do what has been promised. To an American Army company commander, it indicates unwillingness to assume personal accountability. When an Afghan National Army platoon leader agrees to meet for a joint mission at 6:00 a.m. and arrives at 7:00, the platoon leader, with the magic expression "Insha'Allah," invokes the divine, and the American commander chalks it up to incompetence (see also chapter 3 by Eliason in this volume).

American troops also have their own verbal magic. In a conversation one night with a crusty sergeant first class (SFC) who had served multiple deployments, I mentioned that I had gone off base with a patrol convoy and that we had returned via the wadi. *Wadi* is an Afghan word for a waterway that is usually fairly dry. Except during the spring snow runoff, Afghan wadis double as large-vehicle highways far easier to navigate than the small dirt roads designed for carts, donkeys, and motorcycles. The sergeant stopped me: "Ma'am, we don't use that word anymore. We say *arroyo*," he paused, "because shit happens in the wadi." The words one uses come with their own magic. Prayers before missions, unit chants, traditional songs that a unit plays either before or during a dangerous mission offer their own form of protection.

Military personnel don't always bring all the protection they need from home. Sometimes they acquire it "downrange." Chip, a major in the Army Reserves, had deployed to Afghanistan three different times. He recalled that his second deployment in the volatile east of Afghanistan, where he was training members of the Afghan National Police, came with the most danger. After a frightening encounter with insurgents, his commander took him aside, reached into his pocket, took out a set of Islamic prayer beads, handed them to Chip, and said, "I think you need these more than I do" (Chip 2010). Every foray after that Chip took the beads with him, and he credits them with keeping him safe. Neither Chip nor his commanding officer is a Muslim, neither knows that the beads are called *tasbeh*, and neither can recite the ninety-nine names of Allah that a typical believer reviews when "saying" the beads. The magic of the beads came from the fact that they were a gift, and the accumulation of days that he carried them without injury added to their power.

Like Chip, other members of the military attribute the power of protection to a religious object of a faith that they may not share. Captain Marc Motyleski was serving in Bagdad, working with a civil affairs unit outside the base and outside the protected Green Zone, when he found himself trapped and under direct fire:

> While on my second deployment in Iraq, I was conducting an assessment of a substation, which was at the bottom of a small hill. Then all of a sudden we were taking small-arms fire. While peeking our heads out to look around for ways to get out, we received continuous fire. Our battalion commander came on the scene and called in some Apache helicopters. And, of course, the shooting stopped once they came on station. After I came back to base, my friend, Ray Wallace, gave me the LDS [Latter-day Saints] dog tag, saying, "You need this more than I do." He was at the time at a desk job on base. This story ends sadly, however, when my friend was killed in Afghanistan in 2010. (2010)

One side of the silver oval of the LDS dog tag read, "In case of need notify LDS Chaplain or member" above an image of the famous Mormon temple in Salt Lake City. The reverse read, "I am a member of the Church of Jesus Christ of the Latter Day Saints." Motyleski was not a Mormon nor did he intend to become one, but he attached the amulet to his key chain and never again went in harm's way without it. His friend had given him an object that attested to his own faith and with the gift came the object's power, regardless of the faith of the receiver. The power resides with the object, a material thing that can be given away. Protective objects are, in the eyes of their owners, special because of their gift status, not so much in the Maussian sense of gift as a commodity implicitly demanding a reciprocal exchange (Mauss 2004) but more akin to the giving of a medicine bundle intended simply to protect the receiver. Both of these stories are what I call "You need this more than I do" stories, and they illustrate not just the power of an object to protect but also the bonds formed between members of the military. Ray Wallace did not deliberately lay down his life for his friend by handing over his amulet, but his generosity, perhaps a fatal one, suggests that he would have made such a sacrifice.

THE GIFT

In December 2011, I had returned from Kabul and was in Bagram, the way station for travelers going east, west, and north in Afghanistan. Thousands of transients were trying to catch a flight that would take them home for Christmas, return them from a Thanksgiving break, or simply allow them to find their way back to their home base after duties had taken them elsewhere in Afghanistan. Those coming from or going on R & R always had priority; the rest of us sometimes spent days in "Trashcanistan," as disgruntled soldiers called Bagram, to catch a flight. When I wasn't waiting in the cement terminal on one of the metal chairs or drinking coffee at the USO across the street, I was either crashing in temporary billeting or eating at one of the handful of chow halls at Bagram. One night I was having dinner at the Dragon Dining Facility just off Disney Avenue and started a conversation with the Air Force officer seated next to me. When he wasn't in Afghanistan, this officer taught English at the Air Force Academy, and since in a previous life I had taught English at the Naval Academy in Annapolis, we swapped stories about service academies. We discussed our scholarly interests, and I told him that I was currently collecting information on the good luck charms that soldiers and civilians bring to war with them. All of these discussions end the same: the person I'm speaking with pulls out his protective token to show me, and I, in turn, show him a cross my father took to war with him and a Catholic medal a friend sent with me as I headed off to Afghanistan. She had given me another from her collection of medals when I had gone to Iraq in 2008–2009, and since I came back from that encounter in one piece, I gratefully accepted her medal for the longer deployment.

An Air Force master sergeant (MSG) seated across the table, silent until then, pulled out a commemorative coin that he had picked up at the Pentagon. One side of the coin reassured the owner that "God Will Protect," and the other displayed the crest of a unit in which he had served. It was worn, and he told me that he always carried it with him when he deployed. Most military units mint their own "challenge coins" and give them out as tokens of respect and appreciation to individuals who have been of help. The tradition associated with these coins and the explanation of their name come from a common barroom practice. One person pulls out his coin, slaps it on the bar, and challenges everybody else to produce his or her coin. The person who fails to produce a coin must buy a round of drinks. If all present can slap the counter with a coin, the challenger must buy a round. Although all challenge coins are gifts, they are not all taken on deployment as tokens of good luck.

The MSG who had pulled out his good luck coin said, "But what I really always find are hearts—scraps of paper in the form of hearts, pieces of tin in the shape of hearts, and heart-shaped stones." I replied, "You clearly see the trees for the forest, like those people who can always spot the four-leaf clovers in the grass." We soon concluded our cordial conversation, and I was gathering my paper plate and plastic cutlery to take to the trash when the MSG reached into his pocket and pulled out a small heart-shaped rock and handed it to me. On every mission after that I carried the rock in a small pouch on the outside of my body armor.

Amulets not only have power; they also acquire it. Every time one goes "outside the wire" and comes back without incident, the more power the amulet is perceived to have. Hostile situations demand adjustment. Anyone going into dangerous environments on a regular basis must put aside fear. Nancy (who does not want her last name used) spoke candidly about the adjustments she went through during her six-month deployment to Iraq in 2008–2009:

> I became very superstitious about the items that I wore. I had a system every morning in which I put various items in certain pockets of my ACU [Army combat uniform]. These items included an angel figure in the lower-left pocket and a particular coin (with a religious saying) in my upper-right pocket. Other items, such as my knife and my ID, had to go into particular pockets as well. It was very OCD, which is not like me. I knew that it made no rational sense, but my thought was this was the way I had done things so far, and it had kept me from harm—and I didn't want to jinx myself by messing up the routine. (2009)

Before she deployed, Nancy's fiancé gave her a gold rosary, which she never took off. Then her Iraqi interpreter gave her a pendant in the shape of Iraq with a cross in the middle, and she added that to the dog tags and rosary around her neck. Then her teammate and friend, Jean, bought matching cartouches, hieroglyphs that in ancient Egypt had been fashioned to protect the royal wearer from evil spirits after death. Jean had these inscribed with the Arabic word *habibiti* or "dear friend." Jean added her cartouche to the angel necklace that her son had given her before she deployed, and Nancy added hers to the amulets she already wore around her neck (figure 1.2).

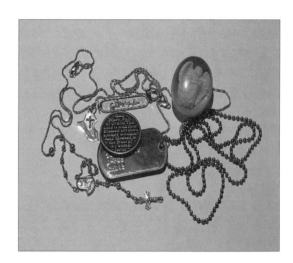

1.2. Amulets worn by US
soldier Nancy during her
deployment in Iraq.

During deployment, the loss of a prized charm can be very disturbing. Once Jean left the angel necklace that her son had given her in the
bathroom. She was very upset because she considered the loss a bad omen.
According to Nancy, "She put a sign in the bathroom about the necklace
being special to her because it was from her son, and surprisingly, someone
brought it back to her. I think because there was this inherent understanding that these types of things were sacred" (2009). Sometimes amulets continue to serve their purpose after deployment. Everyone who deploys passes
through a period of readjustment and assimilation upon return, and during this time, the protective token(s) may help ease the adjustment. Nancy
notes how reluctant she was to be without all those items that she had worn
continuously during deployment: "When I came back, I wore these combinations of necklaces for a long time because it felt 'wrong' to take them
off. It also felt very odd and wrong to take off my dog tags because these
too became part of my 'talisman' even though all of this 'bling' was a crazy
amount of jewelry around my neck."

STUFFED ANIMALS

Girlfriends and boyfriends left at home send soldiers and Marines off
with their stuffed animals, or one appears in a mailed care package. If favored
possessions, these stuffed animals will carry the scent of the loved one. In
fact, some wives and girlfriends spray the stuffed animals with their favorite
cologne so that the object comes with the scent of the giver. There may be

more than mere sentimentality to this practice. Christopher Peterson, who teaches psychology at the University of Michigan, took a poll of his class of 250 asking undergrads who had come to college with a stuffed animal to raise their hands. Over 80 percent of the females in his class admitted to bringing a stuffed animal to college with them, whereas less than 10 percent of the males had. Peterson goes on to speculate that the oxytocin, known as "the cuddle hormone," stimulated by humans touching other humans and petting animals may, in fact, be generated by contact with the soft facsimile of an animal (2010). Far more clinical studies of oxytocin have been conducted on women than on men. This is, no doubt, a function of the drug's historical use in stimulating labor. One study, however, in which both men and women participated, showed that oxytocin fostered social affiliation, empathy, and stress reduction in both men and women (Rodrigues et al. 2009).

Although I observed stuffed animals in many tents in Afghanistan, those that were taken on missions as protective amulets were typically small and fit neatly in a pants pocket. Just before one Marine officer was about to deploy to Afghanistan, her five-year-old son gave her his favorite toy, a small green floppy frog. She asked if he was sure he wanted to part with such a precious toy, and he agreed that she could take it for her first month away. Every couple of days during that month she took the frog out of one of the pockets on the leg of her pants, photographed the frog, and e-mailed the picture to her son. He saw his frog on a commercial aircraft chartered by the Department of Defense bound for Kuwait, in her tent in Kuwait passing the time till she could catch a flight to Afghanistan, on the troop transport airplane to Bagram Air Base, in her new makeshift office, in her small rustic sleeping quarters, at the chow hall, and on all of her travels within Afghanistan. After the officer had spent a month downrange, her son was so delighted at having sent his own delegate to accompany his mother that he wanted the frog to keep her company until they both returned. She called his toy her "frog of war," and photographed him throughout Afghanistan.

I met that Marine Corps officer at a four-day meeting at Camp Julian, a US base just outside Kabul where delegates from various ISAFs (International Security Assistance Forces) gathered in January 2010 to recommend to General David Petraeus the type of training program the Army should undertake to prepare female engagement teams, those small, predominantly female units whose task it is to engage local women. The Marine Corps was the first to demonstrate the success of its female engagement teams, originally referring to them as the "lionesses of Iraq," and it has since improved its training program. The current Marine teams, trained and commanded by this major with her son's green stuffed frog in her pants pocket, had a good deal to teach an army eager to implement its own program.

At the top of a steep hill overlooking Camp Julian stood the lovely Queen's Palace, still majestic despite the Soviet attack on the structure in 1979 that left its roof in disarray and its interior ripe for looting. Each of us, at one time or another, snuck out of our meeting to hike up to the Queen's Palace and walk through the abandoned royal dwelling. It was here that the frog of war posed for more photos—on the dusty but elaborate staircase, in the stately meeting rooms, in the royal bathroom, and on the walls surrounding the palace. As the months rolled by, the officer's son got used to the absence of his frog. If he couldn't be with his mother, at least his beloved frog could.

ELITE UNITS: EVEN THE STRONG NEED TO HEDGE THEIR BETS

The uniformed services are by no means uniform. Not only is Army culture quite distinct from Air Force culture, and the Marine Corps emphasis on small-group solidarity quite different from the corporate Navy of which it is a part, but within every branch of the military vibrant subcultures thrive. West Point and Annapolis, for example, borrow many aspects of their culture from the British boys' schools upon which service academies were modeled. Army Special Forces, the "snake eaters" of the Army, operate as small squads often far from headquarters for long periods of time and enjoy freedoms that other units don't. They maintain relaxed grooming standards and modify their uniforms in ways that would draw rebuke from the commander of a regular unit.

Recruits to the elite group of the Marines, scout-snipers, are selected from those who show the most aptitude and score the highest on technical exams, and they endure a grueling nine-week training program. A selectee, called a PIG, short for "professionally instructed gunman," receives a heavy sandbag, called his "baby pig," that he must carry at all times. He learns to maneuver in rough terrain and to practice his "skull dragging" by moving through a terrain with his face to the ground. He learns that the best way to survive in enemy territory is to avoid the typical byways and opt instead for the swamp not taken, affectionately referred to as the "pig pond." He learns to hit targets half a mile in the distance. By the end of the training

1.3. Hog's tooth.

program, roughly one-third of his class will have failed to qualify. Some who attended sniper school report that after several days of sleep deprivation and utter exhaustion, they began to hallucinate. One trainee, while filling out a written assignment, kept brushing off his paper. When others in his class asked what he was doing, he explained, "It's raining on my paper; I'm just brushing off the water." Some classes are so tightly bonded that they experience collective hallucinations. When one exhausted group was out on a forced run, one trainee said to his classmate, "The leaves on the trees are little ninjas," and his fellow Marine responded, "Can you see the pirates in the leaves laughing at them?" (Girth 2010)

Upon graduation, the freshly minted Marine scout-sniper becomes an official HOG, short for "hunter of gunmen," and accepts with pride his traditional "hog's tooth," a sniper bullet through which a hole has been drilled and a necklace of cord attached (figure 1.3).

Tribal in practice, the totemic hog's tooth marks an elite cohort and sets its members apart for life. A distinguished military subculture, scout-snipers can quickly investigate anyone's claims to membership by getting on the network of fellow elite Marines.

The hog's tooth is a mark, according to Girth, "to serve as a reminder of what I'm capable of." He goes on to explain its value. "It was the toughest thing I ever earned. My Harvard degree doesn't even compare." This Marine was returning from a civilian deployment in Iraq and wearing his hog's tooth when he encountered one of the sailors who works the customs tent at Ali Al Salem, the American base outside of Kuwait City that processes Marines, soldiers, and civilians returning from deployment. The vigilant sailor, thinking the hog's tooth might constitute a live round, demanded that this former sniper surrender his hog's tooth. "I refused," he said, "and was seriously ready to turn right around and spend several thousand dollars to fly commercial just to keep it. The sailor saw I was serious and said, 'Okay, just this once'" (Girth 2010). These are the guys for whom the rules are bent.

According to tradition, the hog's tooth represents the bullet intended for its wearer, and as long as he possesses it, it can never be used against him. The hog's tooth holds power like that of the familiar Indo-European amulet with the image of an eye believed to nullify all evil glances that might cause harm. (For more on the evil eye, see Dundes 1992.) Although the hog's tooth signals the accomplishment of the wearer, its potency can be transferred. A Marine scout-sniper who requested anonymity told me that the fourteen-year-old son of one of his former buddies had developed brain cancer. The boy's dream was to someday be a scout-sniper, so the Marine parted with his hog's tooth and sent it to the boy along with a ghillie suit, a piece of camouflage clothing that allows the sniper to blend in with the foliage in his area. Over a year has passed, and the boy is still doing well. This transference of the object from the Marine to the future Marine illustrates the importance of traditional objects that link one generation of warrior with another. Even though the object is a sign of one man's accomplishments, it holds its own magical power that can be used in another war—battling a life-threatening disease.

Some elite units leave all personal medals, wedding rings, and jewelry on base before going out on a mission so as not to carry with them any personal identification that could be used by the enemy were they to be captured. One Italian Special Forces platoon commander who served two tours in Bosnia-Herzegovina, one in Mozambique, two in Iraq, and three in Afghanistan bought a Ranger brand cigar at the PX and smoked it before going out on a dangerous mission. The mission went well, and although not particularly superstitious himself, he kept up the tradition and soon discovered that the magic was contagious. When his men saw him smoking a cigar before a mission, he noticed that they were happy because they believed that the mission would be successful. His Ranger brand of cigars was available in Bosnia-Herzegovina and in Iraq, but when he was stationed in Afghanistan and Mozambique, he had to opt for a new brand, Swisher

Sweet. The switch was fine, and his luck held, he believed, as long as he stuck to the same brand throughout the deployment. His platoon now has a reputation throughout the Italian Special Forces community as a lucky unit; its members have faced close call after close call, but every time they have cheated death. Eleven years have passed without a loss of a single life, so this commander keeps smoking his pre-mission cigars. Who is he to mess with tradition? (Winer 2012)

WHAT MONEY CAN'T BUY

Why would people who don't typically consider themselves superstitious perform a good luck ritual or wear a good luck charm? In today's asymmetric wars Western forces hold the visual advantage, the panoptical advantage, to borrow a term from Jeremy Bentham (1791) by way of Foucault (1977) and Virilio (2000). From visual devices aloft, we can not only spot a child picking a flower in a field but identify what kind of flower it is. Despite this impressive technological superiority, the enemy still has the stealth advantage. When a typical military unit moves, it hops into a convoy of heavily armored vehicles that can exceed ten tons each and come with a price tag of over $500,000, and lumbers noisily off the base. The enemy, on the other hand, zips up and down the roads and pathways in a convoy of motorcycles, two to a vehicle, much faster and quieter than US troops, with the possible exception of a Special Forces unit. Despite our technological advantage, our forces have seen their freedom of movement severely limited both in Iraq and Afghanistan by a small, easy-to-make device, an IED that costs only a few dollars but that can produce impressive destruction. Although war is always plagued by unpredictability, IED's have increased that unpredictability tenfold. In response, the ordinary soldiers hanker for a comforting sense of certainty—if I do this, then that will happen; if I carry this charm with me, then I will be safe. Amulets inoculate the soldier from uncertainty and fear. Like cleaning his rifle, they give him something to do that adds an element of personal control. By putting something on, by carrying a certain image, or by playing a certain song before heading outside the wire, he can create at least the illusion of control, a hedge against the danger.

In *Sex, Culture, and Myth,* Malinowski (1962) points to the prevalence of magic in the face of danger and emphasizes that what we call "superstition" in others is often the very thing that we refer to as "belief" in ourselves (188). He examines the way in which a magical practice can psychologically "lead to a mental integration, to that optimism and confidence in the face of danger which has won to man many a battle with nature or with his human foes" (189). According to Malinowski, religion gives man "mastery of his fate" and science gives him "control of natural forces," but magic offers him "the grip of chance, luck and accident" (265). Malinowski made these observations fifty years ago, and today there is still much we do not know about the brain chemistry associated with personal objects believed by their owners to hold some power. However, two recent studies provide impressive empirical evidence that devices and practices believed to bring good luck can positively affect performance.

In the first study, Lysann Damisch, a German psychologist, gave subjects a putting iron, one ball, and ten tries to sink the ball into the hole as many times as they could. Half of the group was given a golf ball and told that it was the ball that everyone before them had used. The other half was given a ball and was told that it was a "lucky ball." Those in the latter group performed 33 percent better than those in the former group. In a related study, Damisch and her colleagues invited individuals to bring a lucky charm with them. When they arrived, staff members collected the charms and told participants that they were putting them in another room. They were then divided into two groups. The staff returned their lucky charms to the members of one group but not the other. Each group was given a standard memory test. Those who had their lucky charms with them scored better on the test than did the group without them. Taken together, these experiments showed that a magical object increased manual dexterity, memory, puzzle solving, and self-confidence (Damisch, Stoberock, and Mussweiler 2010).

WHAT THEY BRING BACK WITH THEM

Leslie Burkett, who calls herself a "very proud Marine mom," wrote on an online website for Marines and their family members, www.grunt. com, "My Marine son just returned from Iraq two weeks ago. The first thing he wanted to do was go skydiving—which he did. The next thing he did was to get a tattoo that included the names of the fallen Marines in his Company. I was touched that my 'tough Marine' wanted to have such a permanent memory of his fellow Marines that will stay with him no matter where his life leads him" (Burkett 2008). Former Marine commander Colonel Mike Denning joked that after a week home from deployment,

1.4. A "globe and anchor" tattoo.   1.5. A memorial tattoo.

the Marines in his unit would have "a new car, a pregnant wife, and a tattoo" (2012). Some tattoos display the iconic "globe and anchor" emblem of the Corps (figure 1.4).

Others, such as the one shown in figure 1.5, serve as memorials to fallen friends.

Typically, Marines go with a buddy or a group from the same unit to have identical tattoos. Not only do they honor the fellow Marine and friend they lost but, according to Denning (2012), they further cement the bond among the survivors. In soldier lore, the fallen soldier often comes back in future deployments to act as a guardian spirit to lead one of his fellow mates out of harm's way.

Those returning from deployment bring home Afghan souvenir rugs with images of AK-47s, helicopters, and Soviet tanks woven into them. They bring flimsy belly dancing costumes and the hope that a wife or girlfriend will show what the see-through yellow or pink costumes with coins dangling from them look and sound like on a real woman. They bring home crudely polished Afghan lapis lazuli, cloudy emeralds, and mysterious black diamonds. Some bring blue burkas to show a wife what she'd have to wear in public were she an Afghan woman and have her carry on a conversation through the small crocheted grill work in the front, a conversation they never could have with an Afghan woman. Some bring home a sexually transmitted disease because condoms, by the boxful, are available only in the women's latrines. When I questioned why condoms were absent from the men's latrines, an officer explained, "'Cause these guys would be fuckin' each other right and left." Did those in charge determine that a box of condoms in the male latrines might appear to condone homosexual sex? Were they simply in denial?

They come back with the residue of the war in their lungs from the plastic trash that's burned on base. They bring home biceps inflated from hours spent lifting weights because there's nothing better to do and from the vats of powered protein they've downed because they wanted to look good for each other. They return with dry skin from the months of showers in water that's not really clean but that will clean, water with bleach to neutralize the most dangerous parasites. They bring home a new regard for their command—either a seasoned respect or a quiet disdain. Some bring home embarrassment at having spent their days in a cubical at headquarters where the only danger was dying of boredom when their families imagined them humping the Hindu Kush.

Others bring home nightmares and wounds that nobody can see. They bring home photos that document a time and a place that is strangely out of time and out of place, photos that only their buddies understand. Even though such a practice is in violation of the Uniform Code of Military Justice, a few even bring back war trophies that might in future deployments morph into talismans. They bring home the intimate knowledge of what really happened when one of their soldiers died rather than what was reported in an official account, cleaned up to disguise the blunders, then sent up the chain of command. They bring home the conviction that it just doesn't pay to look too deeply into things or to remember with too much precision. They bring back the expectation that things at home will be just as they left them, that they will slip smoothly back into the slot that was left empty when they deployed. And after their long stint away, they bring home the small tokens they have carried throughout their deployment, the special items that, along with their skill and their alertness, helped them survive. They will put them aside until they need them for the next deployment, or until they give them to the next generation to go into harm's way.

WORKS CITED

Bentham, Jeremy. 1791. *Panopticon Penitentiary House.* London: T. Payne.

Burkett, Leslie. 2008. *AmericanCourage Newsletter* #174, May 15. http://www.grunt.com/corps/newsletter/3198/.

Chip (last name withheld). 2010. Conversation with the author, November 4, Kansas City.

Damisch, Lysann, Barbara Stoberock, and T. Mussweiler. Jul 2010. "Keep Your Fingers Crossed! How Superstition Improves Performance." *Psychological Science* 21 (7): 1014–20. http://dx.doi.org/10.1177/0956797610372631. Medline:20511389.

Denning, Mike. 2012. Phone interview, January 7.

Dundes, Alan. 1992. *The Evil Eye: A Casebook.* Madison: University of Wisconsin Press.

Foucault, Michel. 1977. *Discipline and Punish: The Birth of the Prison.* New York: Random House.

Girth (requested that he be cited by his nickname). 2010. Interview.

Malinowski, Bronislaw. 1962. *Sex, Culture, and Myth.* New York: Harcourt Brace.

Mauss, Marcel. 2004. *The Gift.* Oxford: Routledge.

Motyleski, Marc. 2010. Conversation with the author on December 10, 2010, and follow up e-mail on January 12, 2012.

Nancy (last name withheld per request). 2009. Conversation with the author on January 4, 2009, in Northern Iraq and follow-up e-mail posted to the author on January 12, 2012.

O'Brien, Tim. 1990. *The Things They Carried.* New York: Houghton Mifflin.

Peterson, Christopher. 2010. "Did You Bring a Stuffed Animal to College?" Retrieved from http://www.psychologytoday.com/blog/the-good-life/201011/did-you-bring-stuffed-animal-college. November 4.

Rodrigues, Sarina, Laura Saslow, Natalia Garcia, Oliver John, and Dacher Keltner. 15 Dec. 2009. "Oxytocin Receptor Genetic Variation Relates to Empathy and Stress Reactivity in Humans." *Proceedings of the National Academy of Sciences of the United States of America* 109. http://www.ncbi.nlm.nih.gov/pmc/articles/PMC2795557/?tool=pmcentrez.

Saiyasombut, Saksith. 2011. "Keep Your Talismans Close, Boys." *Bangkok Post.* Retrieved from http://asiancorrespondent.com/48727/black-khmer-magic-a-threat-to-the-thai-army/.

Virilio, Paul. 2000. *The Strategy of Deception.* Trans. Chris Turner. London: Verso.

Winer, Alison. 2012. E-mail to the author, January 7.

2

Know Thy Enemy
Camel Spider Stories among US Troops in the Middle East

Justin M. Oswald

> *I would think that a camel spider seems always to be on attack mode.*
>
> —Matt, a US Air Force veteran who served in the Middle East

> *Rumors express and gratify the emotional needs of the community in much the same way as day dreams and fantasy fulfill the needs of the individual.*
>
> —Robert Knapp

In a classic study of social psychology written toward the end of World War II, Robert Knapp stated that rumors thrive in conditions of "social duress" and that war, which "focuses and intensifies the emotional life of the public," is an especially fertile ground for their circulation (1944, 22). More recent studies of this folklore genre have confirmed this observation; thus, it is hardly an exaggeration when a prominent Israeli folklorist claims that "wars have always been great hotbeds of rumors" (Hasan-Rokem 2005, 45). Since rumors are in general responses to social anxiety, the hotbeds can exist both on the battlefield and in civilian populations; but they are perhaps a little hotter in the former, where soldiers are faced every day with life-threatening situations.

Knapp identified three common types of wartime rumors: pipe dreams, wedge rumors, and bogie rumors. The first, he claimed, satisfied the emotional need for wish fulfillment, the second that of aggression, and the third that of fear. Not surprisingly, combat zones are especially rife with "bogie"

DOI: 10.7330/9780874219043.c02

rumors, including fearful, sometimes fanciful, tales about the enemy and its terrain. Such fear fantasies, which emerge naturally during wartime, may range from mere pessimistic tales to all-out panic rumors about espionage, atrocities, or an impending attack (1944, 23–24). We can see the same process that Knapp described operating today, in the appearance of bogie rumors about the war on terror. The online "urban legend" site Snopes. com, for example, has recently reported rumors about al-Qaeda poisoning cans of Coca-Cola with anthrax or arsenic, and about terrorists stealing rental trucks that they are planning to use in the ongoing attack on America (Mikkelson and Mikkelson n.d.). Such bogie rumors, which used to spread by word of mouth, now circulate much more quickly over the Internet.

The popularity of rumors—and of frightening stories in general—in war zones may be intensified by the fact that soldiers there are not only in harm's way but physically and psychologically removed from familiar surroundings. As human beings, we depend on routine and mundane practices to bring order, consciously or subconsciously, into our daily lives. The inability to maintain our daily habits may engender feelings of alienation and cause us to act outside our usual spectrum of behavior (Fine and Ellis 2010, 2). When we find ourselves in a new physical environment, we may experience "the phenomenon of displacement" (Casey 1993, xiv), and this displacement can alter our perception of self and surroundings, thus allowing new forms of thinking, including rumors, to emerge.

In this essay I examine various personal narratives—from simple memorates to more elaborate anecdotes to full-fledged rumors—that are current among US military personnel in Iraq, Afghanistan, and Saudi Arabia, and that concern the fearful characteristics of a local creature whose common name, among soldiers, is "camel spider." Camel spiders are described as seeking out soldiers at night and inflicting painful damage through their use of aggression, speed, and stealth. New arrivals to these countries are warned about the spiders' lethal nature, and the stories grow in transmission until the creatures take on an aura of mythical evil. The narratives function, therefore, partly to initiate newcomers and partly to control them, but also—as I'll demonstrate—as a kind of bonding ritual and coping mechanism. Before they are in the Middle East for very long, soldiers incorporate camel spider stories into their belief systems, making them part of a common military worldview.

In *Raising the Devil*, Bill Ellis calls legends "part of a 'belief-language' that helps individuals make sense of disorienting and stressful experiences"

(2000, 4). As I will show in this chapter, anecdotes and rumors about camel spiders perform a similar function among deployed US soldiers. I'll also show how soldiers' staging of "death matches" between camel spiders (or between them and scorpions) gives them control over an imaginary battlefield and therefore allows them to be more effective in a hostile environment.

Most of the stories I analyze here were told to me by three male veterans of Middle East deployments. Andrew Han served with the US Army in Iraq in 2003 and 2004. Matt was with the US Air Force from October 1997 to June 2008, and served in Saudi Arabia. My brother Christopher Oswald was with the US Marines from September 2002 to September 2006, and served in Afghanistan and Iraq. All three of them served in the Middle East during ongoing conflicts and all three had seen camel spiders, as well as heard rumors and stories. Even though I have neither been enlisted in the military nor seen a camel spider, my duties within the Department of Defense have allowed me to hear similar narratives.

"WARNING: NOT FOR THE SQUEAMISH"

To biologists, camel spiders are known as solifuges, and they belong to the Order Solifugae, which consists of 12 families encompassing 153 genera. So far, over 900 species have been described worldwide (El-Hennawy 1990, 20–27). They are found not only in the Middle East, but also in other hot environments around the world (Savory 1928, 328), and they have enjoyed an unsavory reputation for centuries. The first-century Roman author Aelian, in his *On the Nature of Animals*, created the rumor that a part of Ethiopia was rendered "completely deserted" by an invasion of solifuges and scorpions, while the eighteenth-century zoologist Anton August Heinrich Lichtenstein proposed that solifuges, rather than mice, were the creatures responsible for plaguing the Philistines, as recorded in the first book of Samuel (Punzo 1998, 2–3). The creatures are known by a variety of aliases, including wind scorpions (Bernard 1897), sun spiders (Pocock 1898), hair-cutters (Lawrence 1955), fat-eaters (Punzo 1998), poison-spiders (Cloudsley-Thompson 1977), and deer killers (Walker 2004). The vernacular nicknames are more colorful than accurate, for technically camel spiders are neither spiders nor scorpions.

But while solifuges are not biologically spiders, physically they look like spiders (figure 2.1), and psychologically they evoke the same sense of apprehension that spiders do.

2.1. Close-up of a camel spider.

Liberally speaking, the fear that soldiers exhibit toward the camel spider may be reasonably classified as a "faux" arachnophobia. And since the fear of arachnids is among the most common of human phobias, many soldiers—even before they hear their first scare story—are already predisposed to be wary of them. In writing about phobias among Vietnam War soldiers, Monte Gulzow and Carol Mitchell show that displaced soldiers with preexisting fears were more likely than others to believe the stories and rumors pertaining to their fears, especially when their lives were at stake (1980, 315). The same principle applies to the current context. In the Middle East, a soldier who already fears insects or spiders and is placed in a strange environment may believe anything fearful he or she hears about camel spiders. Given how common those fears are in the general population, it's not surprising that fear stories about solifuges are widely believed.

The known facts about these creatures are not very reassuring. In the sober words of biologist Fred Punzo, they are "solitary, highly aggressive, and do not tolerate similar types of communal aggregations. No species of Solifugae are known to live in groups"; they are also "known for their pugnacity and ferocious fighting behavior" (1998, 117, 183). And they are fast. In Iraq, rumormongering has credited them with a top speed of twenty-five miles an hour, plus the ability to leap six feet in the air (Walker 2004). This is an exaggeration, as their actual "burst sprint capability" is only ten miles per hour. But for someone afraid of spiders, that's fast enough—and it does make them "the fastest of all invertebrates" (Guinness World Records 2009, 41). Camel spiders can also deliver a painful bite.

Given these known facts, the US military has taken steps to alert any personnel bound for the Middle East to the potential hazards of encountering camel spiders. When Andrew was bound for Iraq in 2003, the Army gave him his first admonition as part of a bioterrorism briefing. He recalls the warning. "They were telling us about what we might find overseas and it's kind of funny because they gave it to us as part of the bioterrorism side of the lecture . . . They said that we would see camel spiders in Iraq and they were *scary* so just leave them alone. When we were about to deploy they were telling us that they're really fast, really sneaky, and they . . . they're poisonous and they can kill you if they bite you" (Han 2011).

Field commanders also take the creatures seriously, and they sometimes pass warnings to soldiers in the field via e-mail. Below is an example of such a warning, contained in an e-mail forwarded to me from a colonel in my office while we were stationed at Bolling Air Force Base; I was no doubt one of many who received the warning over the years. Attached to the e-mail was a picture of two camel spiders; one gripping the other with its powerful jaws.

> Sent: Thursday, April 08, 2004 5:05 PM
> Subject: UNCLAS//N05720//PA/SPIDER
> ICK!!!!!!!!
> UNCLASSIFIED
> Force protection isn't only about a human threat.
> From someone stationed in Baghdad. He was recently bitten by a camel spider which was hiding in his sleeping bag. I thought you'd like to see what a camel spider looks like. It'll give you a better idea of what our troops are dealing with. Enclosed is a picture of his friend holding up two spiders.
>
> Warning: not for the squeamish! (figure 2.2)

In this e-mail, the soldiers' attention is drawn to the subject line—SPIDER—and then to the expected response, the ick!!!!!!!! of disgust and fear (note the string of exclamation points). The message itself describes camel spiders as a threat to our troops just as dangerous as the human one. The e-mail circulated widely in Iraq and within a month was picked up by *National Geographic Online News*. The report by this normally sober news source included a detail that moved the story about "threat to our troops"

2.2. A US Iraqi war veteran holds a pair of camel spiders. Their size is exaggerated due to spatial distortion.

from ominous to implausible: "A photo of two huge spiders," ran the report, "each the size of a man's calf, was accompanied by an alarming note. The sender said his or her friend—or friend of a friend—knew a soldier stationed in Iraq who had said that these spiders could inject a sleeping soldier with anesthetic, then chew out a chunk of flesh" (Walker 2004). Experts later revealed that the size of the camel spiders in the photograph was exaggerated due to spatial distortion and that no "chunk of flesh" was eaten, but the psychological effect of these rumors remained strong because the perceived possibility of harm was very real.

The camel spiders' behavioral traits lead them to seek out soldiers by both day and night, or so the rumors say. By day, seeking relief from the Middle Eastern sun, the creature is drawn not to the soldier but to his shadow; in fact, the Latinate *solifuge* means "one who flees from the sun." By night it is drawn to light sources—a behavior that, as the Department of Defense's FHPR (Force Health Protection and Readiness) website (February 16, 2011) notes, explains their apparent attraction to flashlights and campfires. Because this makes tents attractive to the camel spiders, soldiers are routinely cautioned to shake out their sleeping bags and tuck in their bug netting every night before retiring.

But the creatures' wiliness can sometimes confound such caution, as Andrew indicates in a story about a camel spider that had entered a well-secured tent at night while he was writing a letter with the aid of his flashlight.

> The only time I ever saw one up close is when . . . it was pretty late at night and I was writing a letter in my bunk, and uh, my bunk was pretty close to the bay door . . . so I had my flashlight on and I saw a *shadow* moving over

my paper so I kinda figured out what it was and I looked up and there's a
freakin' giant camel spider inside my, my *bug* netting . . . just like waving
its arms at me. And I kinda flipped out because, one I'm pretty OCD so I
tuck all the corners of my bug netting in. There really wasn't any space for it
to have gotten into my bunk, but somehow it had managed to, I guess, like
crawl in through a nonexistent hole . . . And kinda sit there chillin' without
me noticing.

The worst scenario is that it had been like on me somewhere without me
noticing it and then had just gotten in my bunk when I got in. I pulled the
side of my bug netting out and kinda tapped the side and it ran out, and
then I was tryin' to step on it, and uh, one of the sergeants came by and saw
me . . . and it was pretty late at night so he was wondering what the hell was
going on.

So he comes over like a giant freakin' linebacker-looking dude. And uh,
I'm like, "Sarge, I got a, I got a giant camel spider in my bunk," and he's like,
"*Oh, shit,*" you know, "we got to take care of it, it might kill somebody." So
uh, we were hunking around and he tried to step on it, and the thing just
shot out . . . past his foot, and he screamed liked a weak, like a little girl. And
that was probably the highlight of my week. (Han 2011)

Most of the camel stories I've collected belong to one of three subsets
of the "bogie" category. They are either infiltration stories, prevention stories, or contender stories. I'll describe the latter two types later in the essay.
The story Andrew tells here is an infiltration story. He asserts that he took
all possible precautions to keep camel spiders from entering the tent, yet
one somehow crawled in through a "nonexistent" hole. The story displays
the soldier's physical and psychological displacement, sitting alone while he
writes a letter home to his family. The camel spider chooses this moment
to reveal itself and proclaim its mastery of the land by having navigated
the bug netting. Thus, like terrorists, camel spiders can infiltrate a secured
space—tent, clothes, or bed—at any time. This gives them a power advantage that the soldier can redress only through "threat neutralization," that is,
by killing the intruder—exactly what Andrew and his sergeant attempt to
do. That it is a terrifying act, an act of desperation, is clear from the fact that
it can reduce even a "giant linebacker dude" to girlish whimpering.

But while on the surface Andrew's story may seem merely frightening,
it has a more positive function as well, in reinforcing the very "force protection" that the intruder seems to threaten. Stories such as this one may in
fact increase general vigilance on the part of deployed soldiers, making them
more alert not only as guardians of their sleeping quarters but as everyday
participants in the combat mission. Camel spider rumors thus may serve

the same kind of protective purpose as the vagina dentata legends discussed by Gulzow and Mitchell. During the Vietnam War these legends, although ludicrous on face value, were used as scare tactics to prevent soldiers from contracting venereal disease from prostitutes (1980, 313). In the war in the Middle East, camel spider stories warn soldiers to be wary of their surroundings, always check their gear and shake it out, or suffer a different form of infection and injury.

FROM FACT TO RUMOR: THE SOLIFUGE AS BLOODSUCKER

Andrew's infiltration story, while it is undeniably scary, is still a straightforward, first-person story—in folklorists' jargon, a personal-experience narrative—without apparently manufactured details. Such is not the case with all camel spider narratives. Many of them—especially those that rely not on personal experience but on hearsay—are creatively embellished to emphasize the animals' fearfulness. This is the case, for example, in the story about the solifuge extracting a "chunk of flesh" from its victim. In stories like this, while the detail may seem believable, we are leaving the realm of fact and entering that of rumor.

But rumors, of course, may be interpreted as facts when they come from a trusted source (such as another soldier) and seem believable (Fine and Ellis 2010, 25). They may spread rapidly due to situational insecurities and the perceived dangers of ignoring them. They may also thrive because there is a lack of accurate and accessible information (Knapp 1944, 32–36). In a Middle Eastern war zone, where reliable information about solifuges is limited, soldiers may be willing to believe even wild stories, and unable or reluctant to separate the verifiable from the merely plausible. There are also psychological reasons for the spread of such rumors. As Knapp notes, they may serve to satisfy a teller's need for exhibitionism, to provide warnings, to create bonds of emotional support, to undermine authority, or to project emotional conflict onto an external source (34).

One rumor that is common enough to take on the characteristics of a local legend is that camel spiders inject a venom or anesthetic into their victims that enables them to suck out "vital juices." This belief is given credence by no less an authority than *The SAS Combat Handbook*, which says that "when not eating its fellows, the camel spider will eat beetles, scorpions and even small lizards at great speed by injecting a venom that dissolves the internal organs of the prey and then sucking out the resulting juices" (Lewis 2001, 139). In the oral repetition of such a detail, the camel spider's thirst for "juices" grows to monstrous proportions, so that it is able to feast unde-tected on the flesh of sleeping soldiers, and to kill not just "small lizards" but giant camels by sucking their blood. Andrew remarks on this particular belief: "What I think I remember reading somewhere was that they will bite and take some of the blood from camels, but I don't know the exact reason why. I've heard stories from the grunts and stuff about how they saw a camel spider like take down a camel by *biting* it on the neck, but I think it's just BS . . . One of those BS Army stories that just gets passed down and exag-gerated as they get told" (Han 2011).

Andrew is rightly skeptical about this idea, yet in telling about it he contributes to its continued circulation. A similar mixture of doubt and cre-dence appears in my brother Chris's comment on the "camel sucking" idea: "I don't think they necessarily eat [camels] . . . They're called camel spiders because I think they hang *around* camels. I think what, what we learned was they don't eat the camels, but they use them almost as like a way to get around like a, like a parasite. You know what I mean? I don't think they eat camels" (Oswald 2009).

Bloodsuckers and parasites. Both concepts evoke the image of some-thing alien to the self latching onto it and extracting vitality. This old and geographically widespread folk motif helps to explain the popularity of vampire legends—from Bram Stoker's *Dracula* to the *Twilight* series—as well as such horror films as *Alien* and the 2011 TV miniseries *Falling Skies*. In the transmission of the "bloodsucking spider" rumor, the biological fact that only one species of solifuge—indigenous to India—is thought to be venomous (Punzo 1998, 3) is less important than the image of para-sitic invasion, whose psychological resonance makes it appear plausible. A related folk belief is that the solifuges are called camel spiders because they "lay their eggs inside a camel's belly" (Walker 2004). In fact, they acquired this name because many species have a humpbacked appearance (Punzo 1998, 3). But the egg-laying detail persists in oral tradition, as another pow-erful image of the camel spider's victim being "taken over" surreptitiously by the invading creature.

The "alien invasion" motif appears literally in a narrative I heard from Chris about a camel spider so "creepy" that he and his fellow Marines had to kill it on the spot. This is an example of what I call a prevention story:

Ah, the only way I can really describe what it looks like, and I'm *not making this up*, is if you've ever seen the movie *Aliens*, those things that latch onto the peoples' faces; it's pretty much what it looked like.

It looked bony. It looked *big*. It *was* big. It was probably the size of a *dish plate*. Maybe a little bit bigger, about like a twelve-inch in diameter dish plate.

And like I said, its legs looked all bony and it scared the shit out of us and we didn't know what to do with it. Ah, I mean, we weren't sure, we didn't really know too much about them other than they're huge as shit. So we *shot* it.

I mean that thing was just . . . It was *creepy*. I mean, we lived fairly . . . it was right after we left our base so, it was probably within fifty meters of where you know, maybe seventy Marines were staying. So, last thing we need is one of those getting in the sleeping bags because there have been other *times*, not necessarily happened to any of my guys, but other Marines and soldiers and guys, where the spiders will actually come into their hooches, at *night*.

I don't know whether they're trying to get warmth or whatever, but they will sleep in the shacks or hooches, or the little pits we make to sleep in, and they're *creepy*. And from what I understand is they do have fangs and some sort of teeth to eat their prey, and yeah, they're *pretty* creepy. (Oswald 2009)

Like Andrew's infiltration story, this prevention story is not a rumor but a personal-experience narrative, and Chris tells it with great animation, emphasizing both the creature's fearful appearance and its invasive capability. He repeats and emphasizes adjectives such as big, bony, and creepy, and he uses imagery to make this camel spider seem truly foreign—in this case, extraterrestrial. He acknowledges that the creature exhibits vampiric qualities when he alludes to the existence of fangs, and when he suggests that the creature "creeps" toward its victims at night, seeking the warmth of a living body to feed upon. Even though the animal is more than half a football field away, it is to prevent that unthinkable outcome—the threat to slumbering Marines—that it must be destroyed before it has an opportunity to attack.

In Chris's story, we hear echoes not just of vampire legends but of two similar, and similarly widespread, folk phobias about the fear of being "consumed" by an alien other. One is the medieval fear of the succubus, a female demon that is said to visit men while they sleep to engage in sexual intercourse; in one interpretation, this figure plays on men's universal fear of castration (Raitt 1980). The other "bogie" motif is the vagina dentata, an ancient fear fantasy that, as I've mentioned, enjoyed a rebirth among soldiers in Vietnam. Again, the coded fear is that of castration (Gulzow and Mitchell 1980, 11). As nocturnal visitors up to no good, camel spiders echo these older images of threat. And like the succubus and the vagina dentata, they cannot be dealt with in a measured or negotiated fashion. They are almost too terrible to contemplate, and certainly too terrible to tolerate. As creatures whose presence fuses arachnophobia and the fear of castration, they must be eliminated. In Chris's blunt assessment, "We didn't know what to do with it . . . so we *shot* it."

REPRESENTATION: THE SOLIFUGE AS TERRORIST

But if camel spiders evoke traditional fears like arachnophobia and the fear of emasculation, they also evoke a more immediate anxiety: the fear of the local enemy that, like the solifuge, is seen as both deadly and clandestine. In fact, to soldiers stationed in Iraq and Afghanistan, camel spiders are not just like the human enemy; in some ways they represent the enemy. This was the logic behind my former commander's warning, "Force protection isn't only about a human threat." Unpacking the impersonal syntax, what this meant was, "In protecting ourselves, we have more than human threats to worry about." We have solifuges, too—and they must be taken just as seriously as the human enemy.

According to Eric Greene (1995), humans sometimes assign insects human qualities and project their attitudes and beliefs onto them. Even though they are not insects, this point clearly applies to solifuges as well. In the Middle East, soldiers assign them the qualities of the local terrorist enemy, and the traits assigned to the camel spiders are precisely the predatory characteristics of terrorists: "bloodlust and appetite" for the destruction of America and its military (Fine and Ellis 2010, 44).

In perceiving a similarity between terrorists and camel spiders, soldiers note first that both are solitary in nature. While a terrorist cell may certainly attack en masse, it is the individual suicide attack—as difficult to anticipate as it is lethal—that most often worries those in the field. The solifuge is a perfect metaphor for this threat. No narrative portrays the camel spider as working in a group, and many of them emphasize the stealth with which it infiltrates defenses: By the time a soldier realizes its presence, some damage has already been done.

That damage is always the work of a solitary attacker exhibiting fearful amounts of aggression toward US soldiers, seemingly with little regard for its own survival. Thus, fearless aggressiveness is another characteristic that they share with terrorists. Matt, who has seen camel spiders on several occasions and who makes the usual vernacular identification of the solifuge as both "insect" and "arachnid," emphasizes this characteristic in his description. He calls camel spiders "very aggressive, not scared, not frightened, not startled like a normal insect would be . . . [Their behavior is] not the typical behavior of a . . . I would say, of any arachnid. I would think that a camel spider seems always to be on attack mode. Every time I've seen them they've always been pretty aggressive" (2009).

They are also linked to terrorists by their uncanny speed—their ability to escape death or capture by even the best-trained soldier. When Andrew describes his own infiltration experience, he stresses the fact that his sergeant—a seasoned professional—attempted to catch and kill it, to no avail: "I saw that guy step on it and it didn't even look like it took a dent. It just went in a straight line, and yeah that sucker was fast. I don't know exactly what its ground speed would be, but it got from his feet to the other side of the bunk in like a blink of an eye" (Han 2011). Similarly, Matt describes an encounter that he witnessed in which a camel spider seemed to taunt a soldier before it fled: "I've seen one *stand up* to somebody, and then, yeah, *sprint*. I couldn't put a mile per hour on it but I could say pretty fast. Fast enough to scare the *hell* out of you, you know?" (2009).

The camel spider, then, is characterized by its methodical invasion of safe zones, nimbleness in concealing itself from soldiers, capacity to inflict pain without mercy, and an uncanny ability to escape (in most cases) without retribution. Given this profile, it is easy to see how it becomes the natural world's equivalent of the merciless terrorist. It is difficult to assume that soldiers would consciously acknowledge this connection; but it is a powerful one nonetheless, at the subconscious level.

While most camel spider stories are confined to the Middle East, at least one vivid one made an appearance back at a soldier's home residence. In 2008, *CNN Online* published an infiltration story titled, "Stowaway Afghan Spider Kills Family Dog," in which a family was driven out of their home by a "poisonous" spider that had traveled to England in a soldier's luggage. The spider was said to "hiss" at the family dog, which ran whimpering at the sound, and it was assumed this creature caused its death because, according to one family member, it seemed "too much of a coincidence that she died at the same time that we saw the spider." When they searched the Internet, the family discovered that the thing that had killed their dog was a camel spider. The story, which has the hallmarks of an urban legend, adds a grim twist to the usual in-country narrative. In this tale, the terrorist solifuge is not content merely to attack military personnel, but travels to the homeland itself to destroy a family pet. Though this story occurred in England, this incident resonates with US soldiers. It is the ultimate fear in an age of terror: an attack on the heartland by a "secret" invader.

One detail of the CNN story is particularly interesting in its suggestion of the link between solifuges and terrorists: the camel spider supposedly "hissing" at the dog. In fact camel spiders lack the ability to make such a sound (FHPR 2011), yet a detail about hissing (or screaming) appears in other camel spider stories as well. I suggest that in this detail storytellers are, perhaps unconsciously, referencing not the sound of a basically mute animal, but the fierce battle cry of Islamic antagonists, sometimes phonetically represented as "Alalalalala" (Urban Dictionary 2011). Cook (2000) has identified a shrieking battle cry as theologically suspect among Muslims, yet effective as a means of rallying fighters, and no less an antagonist than Osama bin Laden has been shown emitting this cry in an episode of the television series *American Dad* ("Bush Comes to Dinner," aired August 20, 2010). Attributing the equivalent of such a cry to the camel spider, while biologically inaccurate, anthropomorphizes the "terrorist" solifuge in an understandable fashion.

RESPECT: "CAMEL SPIDERS ARE OKAY WITH ME"

The camel spider elicits both fear and respect. In most of the soldiers' stories I have heard, the former element predominates. In one notable exception—a narrative told to me by Matt—the vicious intruder gains the soldier's respect because its ability to humble the toughest soldier provides a corrective chastisement of an arrogant superior. Here the camel spider becomes an ally—a superhero enlisted temporarily to right a wrong. I asked Matt if he had ever seen a camel spider. "Quite a few," he says, and then he continues:

> 1998 was my first deployment in Riyadh, Saudi Arabia, Prince Sultan Air Base, and we lived in tents there. So when I got there every person, every airmen, soldier, sailor, Marine who was assigned there was assigned to live in a tent, and it was a rank structure, so we had, you know, set up by highest ranking was the team tent chief . . . and they were responsible for everything.
>
> We had this one tent chief who I got assigned [who] was, he was a real *jerk* . . . And his actions weren't unnoticed by anyone so, it's kinda deserving, kinda fitting for the moral of this story. But I guess, long story short, he got the biggest room. He was the type of guy that would go in and you know, take your stuff out of the fridge so his stuff could be cold and your stuff just went to the curb side, you know, being an E1-E2 at the time, you know, you

really couldn't say much about it, this E6 Billy Bad-Ass telling you what to do and what's gonna happen and what's not gonna happen.

But one night he was assigned to work day shift, as I was, and half of the tent slept during the day, the other half slept during the night . . . So I slept when he did, and in my little four feet of space, and, you know, compared to his fifteen to twenty feet of space in our Darnel tent . . . [I] woke up to somebody screaming and yelling and running around, and not knowing what's going on because I'm waking up from a dead sleep, and lo and behold, a camel spider was supposed to have been *chewing* on his leg and left its mark on his thigh.

And you know, everybody was taught, when you get in bed you shake your sheets off, when you put your boots on you shake boots out, check your boots and, you know, shake out uniform to make sure nothing creepy crawlies are in there. And I guess he didn't do it before he hopped in bed and he slept with it for a little while, and from what you hear about the camel spiders, when they *bite* you they inject a little bit of venom to where it numbs you to where you can't necessarily feel them, their injection, and then, they start to chew on you, and start to eat, or feed, or bite, or whatever the hell they do.

I know they're *evil looking*, they ain't scared of anything, and it's the only insect I've ever seen, it doesn't scurry it *sprints* . . . And they're pretty vicious, so . . . That was it, but fitting for the story, that dude got his, got what was coming to him and, you know, camel spiders are okay with me. (2009)

Many of the details here are common ones. The nocturnal infiltration, the need to check your clothes and bedding, the camel spider's speed and fearlessness, and the myth about it chewing undetected on a sleeping victim—all of this is familiar from other stories. So is the image of a tough superior—like the "giant linebacker" sergeant in Andrew's story—being terrified by the small intruder. What's new is that the creature attacks not just any soldier, but one who deserves it. Because the "E6 Billy Bad-Ass" (an E6 is a sergeant) is seen as abusing his authority, his punishment at the jaws of a solifuge appears as a "fitting moral" of the story. For the moment, because the sergeant got "what was coming to him," Matt concludes wryly, "Camel spiders are okay with me."

Matt's story portrays the camel spider as a subculture hero to the enlisted men, and a trickster who brings a well-deserved punishment to an abusive superior. Cross-culturally, this is a common role for small, insect-like creatures. "Psychologically," says Cherry, "the role of trickster seems to be that of projecting the insufficiencies of man in his universe onto a small creature who in besting his larger adversaries, permits the satisfactions of an obvious identification to those who recount or listen to these tales" (2003, 551). This analysis is clearly applicable to the "Billy Bad-Ass" story. Thanks to a creature that is normally vilified, the enlisted men's "larger adversary" is "bested," and by listening to the tale of his comeuppance, they receive "satisfaction." For them, the camel spider switches sides momentarily, and the story becomes a more elaborate cautionary tale, reinforcing the usual message about checking your bedding and adding a "moral" about abuse of power. What Galit Hasan-Rokem says about rumors is equally true of this first-person recollection. Such stories are "intricately interwoven with images of control and power. We may view them as manipulative machinations of power . . . [or as] subversive operations undermining power structures" (2005, 31).

SYMBOLIC ACTION: CAMEL SPIDER DEATH MATCHES

In addition to infiltration and prevention stories, soldiers also tell what I term contender stories, about a particular type of symbolic action they call "death matches." These matches are not unique to today's Middle East. British soldiers staged them during World War I. "Solifugae became familiar during the War to our troops in Egypt and the near East, where Galeodes arabs [camel spider] is very common. The soldiers named them 'jerrymanders,' and admired them on account of their extreme ferocity. At one time the men stationed at Aboukir kept pet Solifugae and fought them against each other, like fighting cocks. Each company had its champion, and bets were freely laid on the results of the fights" (Savory 1928, 329–330).

Matt had similar experiences with camel spider death matches: "We used to, you know, catch them, and when we were on post or deployed in the middle of nowhere, catch them and try to fight them against scorpions and stuff, and they'd just rip scorpions apart, and it's pretty funny. We'd dump out an ice chest full of ice and water and watch. Put them in with scorpions, and watch them do that, watch them fight scorpions, so . . . it's pretty good. (2009)

Why have soldiers from World War I to the present used camel spiders in death matches for amusement? This has a twofold answer, both parts involving control of the uncontrollable. The first reason is therapeutic. Promoting a death match—that is, a seemingly chaotic battle over which the "promoter" actually has control—provides symbolic control over the human enemy for which the camel spider and its antagonist are "managed" stand-ins. In addition, the entertainment value of these battles serves as an outlet for aggression, while concentrating on the action simultaneously keeps soldiers' minds engaged in a state of perpetual battle preparedness. By pitting camel spiders against each other or other creatures, soldiers dispel their fear of helplessness on the battlefield as well as fear toward the enemy.

As Michael Jackson has pointed out, for many soldiers, "the imbalance of power on the battlefield [can] be redressed off the battlefield in fantasy, in language, and in symbolic action" (2002, 47). Camel spider death matches are an example of such symbolic action. Through these ritualized mock battles, soldiers may be magically trying to affect the real-life battles in which they face opponents who are less controllable than the caged solifuges. After all, in order to stage a camel spider death match, you first have to catch the symbolic terrorist. This in itself signifies control. Then pitting the creatures against each other—or against scorpions—certifies that you are master of the total terrain, the manager of chaos rather than its victim. Telling contender stories about the matches after they are over may "recertify" their symbolic effectiveness.

But this analysis assumes that, in some sense, soldiers see the captured fighters as representatives not of an external enemy but of themselves. There's an ambiguity here, in that solifuges typically inhabit a symbolic realm of darkness—they are creatures of the "out there," which cannot be controlled—and yet, during death matches, they are "brought in," as it were, to a light-filled arena. Because of this ambiguity, it may be useful to see the death matches as—to bring in a Jungian perspective—ritual manifestations of soldiers' own unconscious, and of their uneasy relationship to their own "trickster" selves.

"The trickster," writes Jung, "is a collective shadow figure, a summation of all the inferior traits of character in individuals" (Jung 1992, 150). If we imagine the qualities of trickster and of terrorist in an unholy alliance, we might see that in camel spider battles, soldiers are permitted to violate the West's conventional rules of engagement and commit acts of terror on a surrogate enemy through role-playing. This cannot be psychologically easy, but it may be necessary. As Fine and Ellis have said, "It is difficult for many Americans to imagine themselves as terrorists, but we continually try, because that role-playing is the only way that we can give meaning to these acts" (2010, 38). Thus, these mock battles may be a way for soldiers to displace their own aggressiveness onto the camel spiders—a way through symbolic enactment for them to conquer not just the adversary, but themselves.

CONCLUSION: FUNCTIONS OF THE STORIES

Camel spider stories, which circulate widely among US troops deployed in the Middle East, conform roughly to what Robert Knapp (1944), in discussing rumors, called a "bogie" or fear-based folk narrative. In this chapter I've suggested that "bogie" stories about camel spiders generally appear in one of three forms. In infiltration stories, the dangerous creature manages to break through a protective perimeter and attack a soldier (often a sleeping soldier) in his own "home." In prevention stories, the camel spider is caught outside of the protective perimeter—the safe zone—and killed before it can do terrible damage. In contender stories, soldiers reenact symbolic battles between the solifuge invader and dangerous scorpions as a way of demonstrating their control over an uncontrollable situation. In all three types, the camel spider remains a figure of immense power, stealth, and mystery—a perfect animal replica of the indigenous human enemy.

What function do such stories serve for those who tell and hear them? The official military response to that question would be that they warn soldiers about a ubiquitous in-country menace. Certainly this is an important primary function of the tales. Since solifuges do inhabit the Middle East, and since they have been known to inflict harm (although not lethal harm) on military personnel, the warnings about shaking out your bedding make good sense. That the initial information about the creatures comes from official sources may enhance the authority of those in command, and even reassure soldiers that "somebody up there" has their best interests at heart. This could have a positive impact on morale. But there are psychological functions of solifuge stories that I believe are more significant than such monitory lessons.

First, they serve as part of soldiers' initiation into the unfamiliar terrain of the Middle East. Like hazing rituals or like the acquisition of military jargon (see Levy's chapter 5 and Gillespie's chapter 6 in this volume), hearing a camel spider story, telling such a story—and, most of all, encountering a camel spider—marks the newcomer to the area as one "in the know," one who has been initiated into the secret fraternity of those acquainted with "the sand" and its lethal denizens.

Second, sharing camel spider stories helps to enhance camaraderie among the initiated. Tales of scary midnight encounters with the insidious creatures, like ghost stories told around a campfire, create bonds of shared experience among newly arrived soldiers who, in many cases, were unknown to each other before deployment. Austin Bay remarks of military slang that it tells the grunt, "Buddy, you ain't in this alone" (2007, 2). Trading camel spider narratives delivers the same message.

Third, in making soldiers vigilant about potential intruders, the stories may put them more generally on high alert. It may increase their overall sensitivity to potential hazards which, in the dangerous atmosphere of the Middle East, may improve their chances of spotting, and therefore thwarting, a human threat. Being battle ready at all times depends, of course, on more than visual attentiveness; but without such attentiveness, nothing else much matters. And if camel spider stories teach one lesson, it is "Keep your eyes open."

Fourth, as the narratives in this chapter make clear, camel spider stories are entertaining. It would be rash to suggest that they are only entertaining, and that details about "chunks of flesh" and "bloodsucking" are trivial diversions. As I have demonstrated above, these have their own psychological benefits. But the entertainment value should not be discounted. A "scary fun" factor may help to dispel the boredom that, as has often been noted (see Mechling's chapter 11 in this volume, for example), can account for a very high percentage of many soldiers' days. And since a bored soldier is seldom a vigilant one, the humor involved in telling and repeating these tales may offset a significant risk to battle readiness.

Fifth, as I noted in my discussion of Matt's "Billy Bad-Ass" story, some camel spider stories can challenge authority, serving as rhetorical correctives to perceived injustices. This may be a relatively rare element in the narrative repertoire, but it is not insignificant. Given the strict hierarchies of the armed forces, resentment of superiors is not an uncommon element of military life, and it is one that surfaces frequently in cadences, songs, jargon, and other folklore (see, for example, Burns's chapter 4 and Tuleja's chapter 12 in this volume). In this context, camel spiders might be seen as impartial "levelers." Since they attack those in power as well as the powerless, they can provide a kind of rough justice when the powerful are unfair.

Finally, because camel spiders are metaphors for the human enemy, stories about the mock battles—the "death matches"—that soldiers make them undergo allow them to act out successful aggression in a play frame, thereby allaying their anxiety about being in harm's way. If Bill Ellis is right in saying that folklore dresses "universal human hates and anxieties in a contemporary cloak" (2003, 57), then it makes sense to say that these soldier-managed confrontations enable them to project their hatred of the enemy onto the solifuges and control a seemingly uncontrollable war through shared fantasy.

The fact that tales about camel spiders are often exaggerated—sometimes to the point of irrational rumor—in no way diminishes their value as coping mechanisms. Indeed, exaggeration may increase their therapeutic potency. In line with David Hufford's observations about supernatural assault lore, we should be less concerned about whether or not camel spider stories are "true" and more with how belief in the animals' "terroristic" powers shape and influence the believing group's behavior and expectations (Hufford 1982, x–xi). It is by appreciating the psychological value of solifuge stories that we gain some understanding of the soldiers who repeat them and the place of the stories in their collective worldview.

WORKS CITED

Anon. 2008. "Stowaway Afghan Spider Kills Family Dog." CNN.com. August 28. http://www.cnn.com/2008/WORLD/europe/08/28/uk.dangerous.spider/index.html.

Bay, Austin. 2007. *"Embrace the Suck": A Pocket Guide to Milspeak*. New York: New Pamphleteer.

Bernard, H. M. 1897. "Wind Scorpions: A Brief Account of the Galeodidae." *Science Progress* 1: 317–43.

Casey, Edward. 1993. *Getting Back into Place: Toward a Renewed Understanding of the Place-World*. Bloomington: Indiana University Press.

Cherry, Ron. 2003. "Insects in Aboriginal Mythologies around the World." In *Oral Literature and Traditions*, ed. Elisabeth Motte-Florac and Jacqueline M.C. Thomas, 543–53. Paris: Peeters.

Cloudsley-Thompson, J. L. 1977. "Adaptational Biology of Solifugae (Solipugida)." *Bulletin* (British Arachnological Society) 4: 61–71.

Cook, M. A. 2000. *Commanding Right and Forbidding Wrong in Islamic Thought*. Cambridge: Cambridge University Press.

El-Hennawy, H. K. 1990. "Key to Solpugid Families (Arachnida: Solpugida)." *Serket* 2 (1): 20–27.

Ellis, Bill. 2000. *Raising the Devil: Satanism, New Religions, and the Media*. Lexington: University Press of Kentucky.

Ellis, Bill. 2003. *Aliens, Ghosts, and Cults: Legends We Live*. Jackson: University Press of Mississippi.

FHPR. 2011. "Camel Spiders." *Force Health Protection and Readiness*. http://fhp.osd.mil/factsheetDetail.jsp?fact=17 (cited on pages 7 and 16).

Fine, Gary Alan, and Bill Ellis. 2010. *The Global Grapevine: Why Rumors of Terrorism, Immigration, and Trade Matter*. New York: Oxford University Press.

Greene, E. S. 1995. "Ethnocategories, Social Intercourse, Fear and Redemption: Comment on Laurent." *Society and Animals: Journal of Human-Animal Studies* 3 (1): 79–88.

Guinness World Records. 2009. *Guinness World Records*. New York: Bantam.

Gulzow, Monte, and Carol Mitchell. 1980. "'Vagina Dentata' and 'Incurable Venereal Disease' Legends from the Viet Nam War." *Western Folklore* 39 (4): 306–16. http://dx.doi.org/10.2307/1499999.

Han, Andrew. 2011. Personal interview, February 21.

Hasan-Rokem, Galit. 2005. "Rumors in Times of War and Cataclysm: A Historical Perspective." In *Rumor Mills*, ed. Gary Allan Fine, Veronique Campion-Vincent, and Chip Heath, 31–52. New Brunswick, NJ: Transaction Publishers.

Hufford, David J. 1982. *The Terror that Comes in the Night: An Experience-Centered Study of Supernatural Assault Traditions*. Philadelphia: University of Pennsylvania Press.

Jackson, Michael. 2002. *The Politics of Storytelling: Violence, Transgression, and Intersubjectivity*. Copenhagen: Museum Tusculanum Press.

Jung, C. G. 1992 [1959]. *Four Archetypes*. 11th ed. Trans. R.F.C. Hull. Princeton, NJ: Princeton University Press.

Knapp, Robert H. 1944. "A Psychology of Rumor." *Public Opinion Quarterly* 8 (1): 22–37. http://dx.doi.org/10.1086/265665.

Lawrence, R. F. 1955. "Solifugae, Scorpions and Pedipalpi." Ed. B. Hanstrom, P. Brinck, and G. Rudebeck. *South African Animal Life* (Lund University Expedition in 1950–1951) 1: 152–162.

Lewis, Jon. 2001. *The SAS Combat Handbook*. Guilford: Lyons Press.

Matt. 2009. Personal interview, October 16.

Mikkelson, Barbara, and David Mikkelson. N.d. "Rumors of War." Snopes. http://snopes.com/rumors/rumors.asp.

Oswald, Christopher. 2009. Personal interview, October 17.

Pocock, R. I. 1898. "On the Nature and Habits of Pliny's Solpuga." *Nature* 57 (1487): 618–20. http://dx.doi.org/10.1038/057618a0.

Punzo, Fred. 1998. *The Biology of Camel-Spiders*. Boston: Kluwer Academic Publishers.

Raitt, Jill. 1980. "The 'Vagina Dentata' and the 'Immaculatus Uterus Divini Fontis.'" *Journal of the American Academy of Religion* 48 (3): 415–31.

Savory, T. H. 1928. *The Biology of Spiders*. London: Sidgwick and Jackson.

Urban Dictionary. 1999. "Alalalala." *Urban Dictionary*. http://www.urbandictionary.com.

Walker, Cameron. 2004. "Camel Spiders: Behind an E-Mail Sensation from Iraq." *National Geographic*, June 29. http://news.nationalgeographic.com/news/2004/06/0629_040629_camelspider.html.

3

"Folk-Folkloristics"
Reflections on American Soldiers' Responses to Afghan Traditional Culture

Eric A. Eliason

Since 1997 I have taught folklore at Brigham Young University. But from 2002 to 2008 I also served in the Utah Army National Guard as a chaplain for the First Battalion of the Nineteenth Special Forces Group (Airborne). In 2004 I took academic leave to serve a tour of duty in Afghanistan. My responsibilities took me and my assistant on scores of helicopter flights all over the area of operations from the nerve center at Bagram Air Base near Kabul to remote firebases in Gardez, Asadabad, Jalalabad, Kowst, and the Pesch Valley, where I worked closely with servicemen and servicewomen of all branches of the military. I also trained and fought together with Afghan National Army and Afghan Security Forces soldiers. Drawing on my academic training and fieldwork experience, I spent countless hours on patrols and as a cultural liaison sitting with village elders discussing their reconstruction needs and working with them for the security of their villages.

Folklorists rarely stop being folklorists even when engaged in other pursuits, especially when these provide rich opportunities to observe, or might require our expertise. In this essay I share some images and thoughts from my Afghanistan experience, focusing on how American soldiers understood, described, and incorporated Afghan folklore. American soldiers working closely with people of very different customs have developed terms for, explanations of, and analyses about the Afghan folkways that impact their lives. The concepts and vocabulary one folk group uses to make sense of another group's folklore constitute a practice I call folk-folkloristics.

DOI: 10.7330/9780874219043.c03   58

This idea has antecedents and constituent parts that have been in use for some time. Understanding that cultural products are judged aesthetically according to patterns and rules particular to the peoples that produce them is the central insight of *ethnopoetics* as introduced by Jerome Rothenberg (1968) and refined by Dennis Tedlock (1972). Likewise, Dan Ben-Amos promoted the term *native genres* for the categories that folk groups come up with themselves to describe their own verbal art (Ben-Amos 1975). Janet Gilmore and Jim Leary note the existence of traditional artists who are "curators of their own traditions" (1987, 13), and the term *community scholar* emerged around the same time among public-sector folklorists to refer to untrained community members who had, on their own, begun to promote and preserve their cultural heritage. These concepts are close to folk-folkloristics but refer to inward-looking cultural reflection on one's own expressive practices.

Gregory Bateson coined the term *schismogenesis* to refer to, among other things, forms of social behavior that emerge between groups as they recurrently react to each other (Bateson 1935; Bateson 2000, 68). William Hugh Jansen's classic essay "The Esoteric-Exoteric Factor in Folklore" (1959) reveals how different folk groups can have very different reactions to the same cultural material, especially when ownership claims are involved. Richard Bauman emphasized that folklore is not bounded and static within groups but is "action" that takes place in social relationships (1972). Dean MacCannell (1992) promoted the idea of "ethnosemiotics" relating to signifying activity—such as taxonomies, lore, literature, ceremonies, and so on—that occurs between groups. The concepts of "syncretism" and "creolization," to describe cultures in contact swapping DNA to produce new expressive culture more than the sum of their parts, have had a growing influence in the field (Baron and Cara 2003). Despite acknowledging the generation of culture along folk group boundaries, none of these scholars explicitly took up the idea that what they saw happening was akin to what they themselves were doing in noticing and commenting on it.

Alan Dundes demonstrated how folklore can be self-referential "folklore about folklore" in his "Metafolklore and Oral Literary Criticism" (1966). Sabina Magliocco (2004) shows how North American neo-pagans have drawn heavily from folklore scholarship as a resource for forming everyday practices and identity. Put together, Dundes's and Magglicco's observations show that folklore can be a kind of folklore studies and vice versa.

The ideas above beat around the bush of folk-folkloristics, but none specifically recognizes that the folk don't merely perform folklore but also analyze and discuss the lore of others. Folklorists have not yet, so far as I am aware, had much to say about how folk groups develop modes of making sense of or fostering another group's traditional arts and practices. I offer this essay and its examples as a step in that direction.

As I write, the Afghan War is ongoing. It began as a response to the September 11, 2001, terrorist attacks on the World Trade Center in New York and the Pentagon. Its initial goals were to destroy the al-Qaeda network that planned these attacks and to overthrow the Taliban government that harbored them. The Taliban government enforced a harsh and puritanical form of Islam on Afghanistan. They closed girls' schools, massacred religious minorities, banned kite flying, imposed the death penalty for many infractions, and sought to destroy any form of expression they deemed non-Islamic regardless of its cultural significance. While allied forces quickly removed the Taliban from power, Talibs retreated to Afghanistan's south and east. From here they have violently campaigned to reassert their control.

My data are several years old at the time of this writing. Much has changed since 2004. I was in country only for a short time and only in the east of Afghanistan. "American soldiers" in this essay refers to the ones I knew during my deployment. My unit had a special mission to work closely with local people and engage with them culturally—a mission that most other units, trained for other purposes, did not have at the time. However, since 2004 overall allied efforts have in fits and starts adopted much of the Special Forces' counterinsurgency/foreign internal defense model of warfare that we practiced (Savage 2010).

"JINGLE" AS FOLK-FOLKLORISTICS

Before its removal from power, the Taliban sought to destroy much of Afghanistan's cultural heritage and folk arts. The most infamous case of this was the demolition of the giant stone Buddhas at Bamiyan (Shah 2001). By 2004, liberation had allowed for an Afghan expressive culture reflowering that influenced the practices of Americans working to help rebuild and provide security for a country ravaged by decades of bloody conflict. The resurgence in Afghanistan of what allied soldiers call "jingle" is a representative example of this.

"Jingle truck" is US soldiers' folk speech for what Afghans simply call a truck (figure 3.1).

3.1. Jingle truck.

Virtually every commercial truck in Afghanistan is heavily decorated (Hallett and Hallett 1972; Centlivres-Dumont 1976; Khan 1975; cf. Kirkpatrick and Bubriski 1994), making them resemble what American folklorists call "art cars," or vehicles modified for striking visual aesthetic effect rather than for performance (Blank 2001, 2007; Dregni and

Godollei 2009). Jingle trucks' colorful extravagance helps enliven what can be a bleak landscape. Some of the designs painted on the trucks are said to be charms to protect against the dangers of the road. Jingle also advertises to potential customers that truck owners are such successful businessmen they can afford thousands of dollars for artisans to fashion and paint elaborate bodywork. "Trust me, I am powerful enough to safely deliver your goods in this dangerous country," the jingle truck seems to boast, in the same way a peacock's showy and burdensome feathers tell a potential mate that he is healthy and vigorous (Dutton 2010, 136–138).

Jingle trucks are invaluable to the war effort, transporting much of the material needed to support operations. Soldiers interact with them and their drivers regularly in the course of their logistical work. The importance of jingle trucks was underscored when supply personnel at Bagram Air Base contracted a nonjingle refrigerated truck to drive the eighty or so miles of boulder-strewn dirt tracks from Kabul to Camp Blessing firebase in the Pesch Valley with a load of morale-boosting ice cream bars. The truck was so slow over the torturous terrain and so obviously out of place that it attracted insurgent attention. The roadside bomb they set blasted a hole through the undercarriage, taking out the refrigeration unit and half the ice cream bars. The shaken but intrepid Afghan driver pressed bravely on to our firebase to make his delivery schedule, but our remaining treats had all melted. Our firebase freezer made the bars solid again, but since melting separated them into their constituent unappealing gums and fillers every bite was a reminder of the importance of jingle trucks.

The term *jingle* specifically refers to the dangling metal chains that often skirt an Afghan truck's undercarriage. But allied soldiers have expanded the term to refer to all kinds of colorful, personalized, but traditionally patterned aesthetic adornments. American soldiers speak of "jingle cars," "jingle bikes," "jingle tractors," "jingle gates" (see figures 3.3, 3.4, 3.5, and 3.6), and even "jingle girls" (see figure 3.2).

They talk of objects being "jinglefied" and are generally amused and delighted by what they see as an Afghan tendency for "jinglefication." "Jingle" and its grammatical variants are American soldiers' folk-folkloristic etic term for particular expressions of Afghan material culture. Folklorists are not alone in being able to develop an analytical descriptive vocabulary for another group's lore.

3.2. Jingle girls.

3.3. A jingle car.

3.4. A jingle bike.

3.5. A jingle tractor.

3.6. Jingle gates.

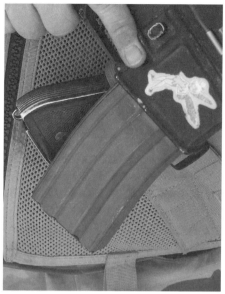

Thanks to Dave Butow of *US News and World Report* for this photograph.

3.7. A Bollywood starlet on an   3.8. Tinkerbell on an American soldier's
Afghan soldier's rifle stock.   weapon.

JINGLE RIFLES

The jingle variety perhaps most relevant to soldiers' experience is the tendency of Afghan Security Forces (ASF) to jinglefy the butt stocks of their AK-47s with colorful translucent tape over photographs and other images. Demonstrating universal young male interests—and perhaps influence from American soldiers' penchant for pinups in wars past and present—Bollywood starlets are a favorite ASF jingle rifle motif (figure 3.7). My friends in the ASF explained that such decoration would have been considered pornography punishable by death under the Taliban. The freedom to jingle is one of the things these soldiers fight for.

Lieutenant Colonel Marcus Custer, then commander of the First Battalion, Nineteenth Special Forces, provides some folk-folkloristic analysis similar to comments I heard from a number of soldiers, both officers and enlisted.

> To me, it seems absurd to decorate an AK-47 like that. It's not my or the US soldier's style. But if you understand Afghan culture, it is a display of individualism and shows pride in themselves as soldiers. This country is so barren, any bit of added color does it good and I see why they do it. Afghans

decorate everything important to them—trucks, cars, bikes. Even though it may seem strange to an American, when I see our Afghan guys wrapping that pink and green tape stuff on their butt stocks, I see it as a positive sign of the pride they take in fighting the Taliban and their desire to better their country. (Eliason 2004b)

Despite the line Colonel Custer draws for himself between Afghan and American soldiers' expressive culture, American soldiers have not only developed terms for describing Afghan folkways, some have been drawn into them as well. This is especially true among soldiers whose mission is to work closely with locals. In rifle adornment, some Americans have followed Afghans' lead, attaching beaded shoulder straps or sporting their own favorite female icons. When I asked the Civil Affairs sergeant whose weapon appears in the image below why he put a Tinkerbell sticker on his magazine receiver, he replied simply, "When in Rome . . ." (figure 3.8).

BEARDS, *JAMAS,* AND *PAKOOLS*

Operating in accordance with the *pashtunwali* (Groh 2010), or traditional Pashtun legal code, Special Forces soldiers wear beards and sometimes don traditional Afghan garb for special occasions. Pashtun *jamas* (suits) are ubiquitous in eastern Afghanistan and the Tajik *pakool* (rolled wool caps) have become pan-Afghan symbols of anti-Taliban resistance (figure 3.9).

Afghans show little concern for foreign academic hand wringing about "appropriation" or "authenticity" and tend to express great delight in what they take—and soldiers intend—as signs of solidarity and respect for the authority of local custom.

American soldiers sometimes call jamas "jammies" or "pajamas." To American eyes, native clothing looks like sleepwear and they find it amusing that the sonic similarity of the Pashtun word underscores this observation. The similarity in name and design comes from the fact that the English and Pashtun words have the same Indo-European root. English borrowed *pai jamah* from Hindi in the 1800s in reference to loose clothing tied at the waist and worn by Muslims but adopted by European colonists in South Asia as comfortable sleeping clothes ("pajamas").

3.9. Afghan mullahs (religious leaders) and tribal elders with US Special Forces soldiers, some in Afghan traditional clothing.

RESURGENCE AND REVIVAL: VERNACULAR RELIGION

In 2004, Afghan religious expression had taken some small steps toward becoming a freely chosen commitment rather than the enforced social conformity it was under the Taliban. The Taliban considered *tarwis* charms worn for protection from illness and jinn (genies) gross superstition that must be repressed. In 2004, tarwis mullahs (religious specialists or shamans who make the charms), a number of whom are women, were enjoying increased freedom. They prepare sacred messages written in divinely given script on specially folded paper sewn into cloth pouches worn around the neck or on an armband (figure 3.10).

Most tarwis mullahs, like most Afghans, are illiterate. The written word carries a powerful mysterious force for many. What literate Afghans might see as nonsensical scribblings tarwis believers insist are passages from the Quran, direct revelations, or both. That post- Quranic revelation is anathema in orthodox Islam is one of the reasons the Taliban suppressed it. But to tarwis mullahs, such distinctions from the literate world are outside their recognition; all legitimate revelation is Quran.

3.10. An Afghan girl wearing tarwis charms.

Some US soldiers solicit tarwis mullahs to make them packets for protection. The reasons American soldiers expressed for adopting this practice ranged from wholehearted belief in their power to "It can't hurt" to a desire to show solidarity with their Afghan brothers-in-arms to the wish to have a curious wartime souvenir. Soldiers would reflect with each other on how they squared tarwis with their own secularism, paganism, Christianity, or Judaism. Most felt that regardless of their own beliefs, it was entirely proper to share respect for what their Afghan friends respected, and wear with solemnity, if not belief, a token of how seriously they took their time in Afghanistan (See Burke's chapter 1 in this volume).

Our Afghan friends took tarwis very seriously and were willing to help us procure them only once they were sure we would not trifle with them. They did this out of fear for our safety as much as any desire to protect their traditions from ridicule. Now that I am home and no longer wear mine, I store them on a high shelf in my office so they will always be above my waist as instructed by my tarwis mullah.

THE AFGHAN "SPRAY AND PRAY"

Not all Afghan folkways for which American soldiers have coined terms are ones they want to emulate or even tolerate. There is one practice in particular that American soldiers training the Afghans have vigorously tried to stamp out. This is a traditional Afghan way of war Americans have dubbed the "the Afghan spray and pray." *Pray and spray* is a term that American soldiers have long used to refer to panicked undisciplined fire that one hopes will be effective with the aid of divine intervention. However, the reasons Afghan soldiers give for their version of this practice have nothing to do with panic or lack of discipline, and they actually hope that their shooting will not be purposely effective.

The US military tradition teaches soldiers to fire weapons with intention and precision so as to maximize efficacy and leave as little to chance as possible. A soldier trains to distinguish a legitimate target (an enemy combatant) from illegitimate targets (friendly soldiers and civilians) on the battlefield. Then he or she selects the target that is the most immediate threat, aims at the target through the rifle sight, and fires controlled single shots or short bursts until the target is no longer a threat. Then the soldier begins the target selection, aiming, and firing process again until there are no more threats (Department of the Army 2003).

ASF soldiers resisted being trained in this way. Instead they preferred to look away from their rifle sight or even close their eyes while they fired off long, fully automatic bursts while waving their weapon around in the general direction of the enemy. This horrified their American trainers, who saw it as dangerous to civilians and fellow soldiers, wasteful of ammunition, and unlikely to eliminate enemy threats. It also violated what American soldiers saw as a defining difference between them and their enemy—they fought to protect innocent civilians from terror, while their enemies specifically targeted children and other noncombatants to inflict as much terror as they could.

Range master Sergeant Jim Trusty trained ASF soldiers in marksman-
ship at Camp Blessing in the Pesch Valley in 2003–2004. His explanation
is an example of folk-folkloristics in action: "They believe that shooting this
way absolves them of responsibility. If someone dies, enemy or not, then it
was God's will because they were not aiming. If they were to take aim and
actively try to kill someone who maybe, in God's eyes, did not deserve to
die, they think they will be held responsible in the final judgment." Sergeant
Trusty was quick to explain that this was not lack of bravery on his trainees'
part, but what he considered bad theology.

Sergeant Trusty and I took this issue up with Maseullah, the ASF mullah I was training as an ASF chaplain. We wanted to know what he thought,
and if he might see a solution to the problem. Maseullah knew of this folk
theological belief and of the "spray and pray" practice. He explained how
Afghans avoid presumptuously trying to be in control of anything when
everything is in God's hands. But he came up with the idea of mentioning
in his Friday sermon that protecting the villagers of the Pesch Valley from
Taliban and al-Qaeda attacks was God's work and that, like anything done
before God, it should be done properly and with exactness. This meant
taking care to avoid shooting noncombatants and being justified in carefully directing effective fire at those who would threaten the peace of the
valley. Maseullah talked to his soldiers about this and some improvements
were made, but the "spray and pray" tradition continues to thwart American
training goals.

"INSHA'ALLAH ATTITUDE"

What Americans saw as Afghan fatalism and unwillingness to take personal responsibility also came up in American reactions to the Arabic phrase
that is ubiquitous throughout the Muslim world, "Insha'Allah," or "If God
wills." Afghans routinely attach this phrase to any prediction of, expression of hope for, or declaration of intent about the future. To an Afghan,
leaving it off would be presumptuous and prideful, even blasphemous. In
contrast, American military culture regards taking personal responsibility
for mission accomplishment, scrupulously keeping commitments on time,
refusing to make excuses, and exuding confidence and energy when div-
ing into any task as identity-defining core values. What Afghans see as an
expression of humility in the face of uncertainty and an acknowledgment of
God's sovereignty over all things Americans often take to be a weak-kneed

hesitancy and a preemptive cop-out. In frustration, Americans labeled this the "Insha'Allah attitude." Occasionally, American soldiers might respect the piety it implied, but they also regarded it as one of the main reasons for Afghanistan's backwardness. In American soldiers' folk-folkloristics, "Insha'Allah attitude" as a term of analysis helped them make sense of things they experienced on a regular basis. They used the term not just in reflection but also in developing practices to work with and around it.

Diverging American and Afghan perspectives on the Insha'Allah attitude led to occasional friction. Among the Pashtuns of Kunar Province, Afghanistan, if a builder agrees to construct a wall, standard business practice is to charge by the meter. The customer pays in installments as each section is built. No deadline is stipulated and the wall gets done when it gets done. But when Special Forces team sergeant Randy Derr hired a local work crew to deliver and spread gravel for Camp Blessing's motor pool yard, Derr told the crew work had to be done by a certain date. The foreman replied, "Insha'Allah." Sergeant Derr vigorously countered, "No! Not Insha'Allah. I want insha'foreman! I want you to tell me you will get it done by this day, and then get it done like you said!" The foreman laughed politely but he and his crew members' eyes flitted about as if scanning the sky for imminent lighting bolts. Despite theological differences and friendly chiding, Sergeant Derr and the foreman still managed to complete many projects working together.

MAN-LOVE THURSDAY

One of the most well-developed folk-folkloristic phenomena in Afghanistan is "Man-Love Thursday"—a military folk speech phrase for a weekly joking tradition among English-speaking forces. This tradition comes from the perception that Afghan men customarily seek sex from other men on Thursday so that they can be ritually pure and free from lust on Friday—a day of prayer and devotion for Muslims.

Soon after allied troops arrived in Afghanistan, reports came from British, Canadian, and American military personnel of Afghan same-sex erotic behavior. Coalition troops described being propositioned for sex and spoke of local customs in which young men and boys, *halekon,* were partnered with older men in erotic-romantic relationships (Carter 2010). Allied soldiers reported that Afghans officially consider same-sex erotic behavior wrong, but privately celebrate it. The extreme form of purdah (seclusion of women) practiced in Afghanistan allows young men virtually no opportunity to form friendships or romantic relationships with young women. Afghan men often meet their wives for the first time on their wedding day. Also, unlike in India, where wives come with a dowry, Pashtun young men must pay a bride price of hundreds or even thousands of dollars. This dissuades many from even trying to marry in such a poor country. Afghan homoerotic practices and attitudes may have less to do with participants' sexual orientation than with the unavailability of alternatives.

Non-Afghan soldiers' perception of Afghan same-sex practices can be understood, in part, as a projection of Western constructs of what homosexuality means onto a very different situation. Outsiders may not easily differentiate what is homoerotic from what is not because Afghan men's folkways include open displays of affection associated with gays in America, such as holding hands and exchanging flowers. On several occasions, mullah Maseullah picked flowers for me while we patrolled with our soldiers together. I was delighted, but many Americans felt uncomfortable with such gestures even though they know they do not imply homosexuality in Afghanistan. Some military authors have stressed how essential to effective counterinsurgency it is to get over such qualms and not reject local gestures of deepest friendship and camaraderie when they are offered (Poitras 2009, iii, 29) (figure 3.11).

Many Afghan men would be careful not to ascribe homosexual activity to their own locale. Instead, they may claim it is commonplace elsewhere in Afghanistan. For example, Kandahar, a city thought to be named after Alexander the Great (famous for his man-love, among other things), has a reputation for being especially enthusiastic about man-love and pedophilia (Reynolds 2002). According to an Afghan saying about Kandahar, "Birds fly over the city with one wing covering their butts." That Kandahar also has the reputation as the country's most religiously conservative city underscores the complexity of same-sex dynamics in Afghanistan.

Out of this complex and semisecret background of distinctive cultural practices, the term "Man-Love Thursday" (MLT) emerged as jocular military folk speech concerning host-country men. American perceptions of Afghan practices have led to a weekly tradition among deployed soldiers. On Thursdays, they often joke around about each other's sexual orientation and respond to inadvertent eyebrow-raising situations—such as two men accidentally bumping into each other in the chow line or a friendly wrestling match that ends up in a sexually suggestive tangle—with a comment like, "That's okay. It's Man-Love Thursday" or "Apparently today is your day."

3.11. The author and mullah Maseullah prepare to visit soldiers on remote observations posts near Camp Blessing in the Pesch Valley, Afghanistan, spring 2004.

This tradition may be seen within the larger context of homophobic folk speech that ascribes homosexuality among men to a rival group. Yet it also demonstrates an incorporation of native homosociality into one's own practice—and the development of an analytic concept to describe it—even if only in jest. Although MLT is potentially homophobic in tone, it is not necessarily so. It can be used as a veiled way of reaffirming the strong camaraderie soldiers often feel for each other.

As folk speech that Western soldiers use to conceptualize Afghan traditional practices, MLT is perhaps one of the best examples of the idea of folk-folkloristics as found among American soldiers.

CONCLUSIONS

As US Army Special Forces live, work, and fight side-by-side with Afghans, these two groups cross-fertilize each other's folkways as they cooperate in reconstructing and providing security. In this often intimate context, American soldiers have developed their own folk-folkloristic vocabulary for making sense of Afghan traditional practices.

So, what can be said about Special Forces folk-folkloristics as it was practiced in Afghanistan in 2004? What schools or varieties of professional folklore research does it most resemble? This study is admittedly cursory and the phenomenon it addresses is little developed by those who practice it, but it mostly seems to resemble public-sector folklore in its practicality, interest in enabling cross-cultural communication, and lack of interest in abstract theorization. The sentiments that animate public-sector folklore's funders, if not necessarily its practitioners, are aesthetic appreciation, heritage education, and the ever-popular preservationist impulse. Conversely, Special Forces folk-folkloristics is a streamlined means to an end, one efficient tool among many for mission accomplishment, or it allows for comic relief and releasing the tensions caused by cultural differences.

In ethnographic practice, folklorists often describe folk groups' knowledge systems as corresponding to scientific or humanistic fields of study, and routinely point out that traditional lore often contains valid complex information unknown to "mainstream" or "official" bodies of knowledge. Ethnobotanists see indigenous cultures' use of plants as akin to pharmacology and botany and have thereby developed modern pharmaceuticals from traditionally used medicinal plants (Schultes and von Reis 1995; Davis 1997; Minnis 2000). David Hufford identifies what he calls ethnopsychology at work in supernatural narrative traditions that he finds make better sense of certain psychological phenomena than mainstream psychiatry does (Hufford 1989; see also Walker 1995). Folklorists enamored of TEK, or "traditional ecological knowledge," claim that local tradition bearers can have better insights, and promote more sustainable practices, than college-educated and officially credentialed ecologists (Ingliss 1993; Berkes 1999; Toledo 2002; Eliason 2004a, 2009). Apparently, folklorists have been quick to see the folk as proficient at other people's academic disciplines, but have had little to say yet about how good the folk may be at our own trade.

I very much doubt that the folk-folkloristics phenomenon is unique to Special Forces soldiers in Afghanistan. In fact, I recently had a folklore student tell me about her roommate who had never taken a folklore class, yet kept a notebook filled with transcriptions of jokes her roommates told, took and organized pictures of their apartment's traditional decorations, and compiled the recipes each roommate had brought from home (Felix 2011). Only after she told me this did I tell her about folk-folkloristics. I look forward to learning about many more variations of this bottom-up practice of folklore collection and analysis as it is done by folk groups other than folklorists in other situations.

ACKNOWLEDGMENTS

This essay has been possible thanks to the soldiers of the 19th Special Forces Group (Airborne) of the Utah, Washington, and Florida Army National Guards, the Afghan National Army, and Afghan Security Forces contingent and local villagers surrounding Camp Blessing in the Pesch Valley, Afghanistan. Special thanks are due Simon Bronner, Jay Mechling, Mickey Weems, Tad Tuleja, Ron Fry, Jim Trusty, Randy Derr, Maseullah, Walid, Hamid, Marcus Custer, Jim Leary, and Margaret Mills. Conversations with these fine individuals persuaded me to write this essay. Their input and feedback has been invaluable in shaping its contents.

WORKS CITED

Baron, Robert, and Ana C. Cara. 2003. "Introduction: Creolization and Folklore—Cultural Creativity in Process." *Journal of American Folklore* 116 (459): 4–8. http://dx.doi.org/10.1353/jaf.2003.0002.

Bateson, Gregory. 1935. "Culture Contact and Schismogenesis." *Man* 35 (December): 178–83. http://dx.doi.org/10.2307/2789408.

Bateson, Gregory. 2000. *Steps to an Ecology of the Mind: Collected Essays in Anthropology, Psychiatry, Evolution, and Epistemology*. Chicago: University of Chicago Press.

Bauman, Richard. 1972. "Differential Identity and the Social Base of Folklore." In *Toward New Perspectives in Folklore*, ed. Américo Paredes and Richard Bauman, 31–41. Austin: University of Texas Press. http://dx.doi.org/10.2307/539731.

Ben-Amos, Dan. 1975. "Analytic Categories and Ethnic Genres." In *Folklore Genres*, ed. Dan Ben-Amos, 215–42. Austin: University of Texas Press.

Berkes, Fikret. 1999. *Sacred Ecology: Traditional Ecological Knowledge and Resource Management*. London: Taylor & Francis.

Blank, Harrod. 2001. *Wild Wheels*. 2nd ed. Douglas, AZ: Blank Books.

Blank, Harrod. 2007. *Art Cars: The Artists, the Obsession, the Craft*. 2nd ed. Douglas, AZ: Blank Books.

Carter, Sara A. 2010. "Afghan Sex Practices Concern US, British Forces." *Washington Examiner*, December 20. http://washingtonexaminer.com/news/world/2010/12/afghan-sex-practices-concern-us-british-forces (accessed 6 May 2011).

Centlivres-Dumont, M. 1976. *Popular Art in Afghanistan: Paintings on Trucks, Mosques and Tea Houses*. Graz, Austria: Druck.

Davis, Wade. 1997. *One River*. New York: Simon & Schuster.

Department of the Army. 2003. DA PAM 350-39 *Standards in Weapons Training* (Special Operations Forces). FY03/04.

Dregni, Eric, and Ruthanne Godollei. 2009. *Road Show: Art Cars and the Museum of the Streets*. Golden, CO: Speck Press.

Dundes, Alan. 1966. "Metafolklore and Oral Literary Criticism." *Monist* 50: 505–16.

Dutton, Denis. 2010. *The Art Instinct: Beauty, Pleasure, and Human Evolution*. New York: The Bloomsbury Press.

Eliason, Eric. 2004a. "Foxhunting Folkways under Fire and the Crisis of Traditional Moral Knowledge." *Western Folklore* 63 (1–2): 123–67.

Eliason, Eric. 2004b. Interview with LTC Marcus Custer, commander of 1/19th SFG(A). Bagram Airfield, Afghanistan. Notes in author's possession.

Eliason, Eric. 2009. "Great Plains Coyote Coursing: Biofacts and a New Folkloristic Understanding of Animals." In *Wild Games: Hunting and Fishing Traditions in North America*, ed. Dennis Cutchins and Eric A. Eliason, 25–43. Knoxville: University of Tennessee Press.

Felix, Krysta. 2011. "Apartment Quote Walls." William A. Wilson Folklore Archive, Special Collections, Harold B. Lee Library, Brigham Young University, UT.

Groh, Ty L. 2010. "A Fortress without Walls: Alternative Governance Structures on the Afghanistan-Pakistan Frontier." In *Ungoverned Spaces: Alternatives to State Authority in an Era of Softened Sovereignty*, ed. Anne Clunan and Harold Trinkunas, 95–114. Stanford, CA: Stanford Security Studies (an imprint of Stanford University Press).

Hallett, Judith, and Stanley Hallett, directors. 1972. *Painted Trucks*.

Hufford, David. 1989. *The Terror that Comes in the Night: An Experience-Centered Study of Supernatural Assault Traditions*. Philadelphia: University of Pennsylvania Press.

Ingliss, J. T. 1993. *Traditional Ecological Knowledge: Concepts and Cases*. New York: International Development Research Center.

Jansen, William Hugh. 1959. "The Esoteric-Exoteric Factor in Folklore." In *The Study of Folklore*, ed. Alan Dundes, 43–51. Englewood Cliffs, NJ: Prentice-Hall.

Khan, Hasan Uddin. 1975. "Mobile Shelter in Pakistan." In *Shelter, Sign and Symbol*, ed. Paul Oliver, 183–96. London: Barrie and Jenkins.

Kirkpatrick, Joanna, and K. Bubriski. 1994. "Transports of Delight: Ricksha Art in Bangladesh." *Aramco World* 45 (1): 33–5.

Leary, James P., and Janet Gilmore. 1987. "Cultural Forms, Personal Visions." In *From Hardanger to Harleys: A Survey of Wisconsin Folk Art*, ed. Robert T. Teske, 13–22. Sheboygan, WI: John Michael Kohler Arts Center.

MacCannell, Dean. 1992. *Empty Meeting Grounds: The Tourist Papers*. New York: Routledge. http://dx.doi.org/10.4324/9780203412145.

Magliocco, Sabina. 2004. *Witching Culture: Folklore and Neo-Paganism in America*. Philadelphia: University of Pennsylvania Press.

Minnis, Paul E. 2000. *Ethnobotany: A Reader*. Norman: University of Oklahoma Press.

"Pajamas." Online Etymology Dictionary. http://www.etymonline.com/index. php?term=pajamas (last accessed 21 January 2012).

Poitras, Maurice V. 2009. *Adoptable Afghan Customs or Practices in a Military Operations Environment*. Fort Leavenworth, KS: School for Advanced Military Studies, United States Army Command and General Staff College.

Reynolds, Maura. 2002. "Kandahar's Lightly Veiled Homosexual Habits." *Los Angeles Times*, April 3. http://www.glapn.org/sodomylaws/world/afghanistan/afnews009. htm (last accessed 20 May 2011).

Rothenberg, Jerome. 1968. *Technicians of the Sacred: A Range of Poetries from Africa, American, Asia, Europe, and Oceana*. Berkeley: University of California Press.

Savage, Storm. 2010. "US Army Combined Arms Center: Afghan Counterinsurgency Lessons Brief." http://publicintelligence.net/u-s-army-combined-arms-center-afghan-countersurgency-lessons-brief/ (last accessed 20 May 2011).

Schultes, Richard Evans, and Siri von Reis. 1995. *Ethnobotany: Evolution of a Discipline*. Portland, OR: Timber Press.

Shah, Amir. 2001. "Taliban Destroys Ancient Buddhist Relics." *The Independent*, March 3. http://www.independent.co.uk/news.world/asia/taliban-destroy-ancient-buddhist-relics-694425.html (last accessed 6 May 2011).

Tedlock, Dennis. 1972. *Finding the Center: Narrative Poetry of the Zuñi Indians*. New York: Dial Press.

Toledo, V. M. 2002. "Ethnoecology: A Conceptual Framework for the Study of Indigenous Knowledge of Nature." In *Ethnobiology and Biocultural Diversity*, ed. John R. Stepp and Associates. Athens, GA: International Society of Ethnobiology.

Walker, Barbara. 1995. *Out of the Ordinary: Folklore and the Supernatural*. Logan: Utah State University Press.

Part II
Sounding Off

4

Where Is Jody Now?
Reconsidering Military Marching Chants

Richard Allen Burns

In the spring of 1944, an African American US soldier named Willie Duckworth was on detached service at Fort Slocum, a provisional training center in New York State. To motivate his recruits to march in step during close-order drill, Duckworth used the catchy, rhythmic call-and-response chants that we know today as marching cadences or cadence calls. After the war, the Department of Defense distributed the cadences in a publication edited by Bernard Lentz (1955), and they became generally known as "Duckworth chants" (Lineberry 2003). Literally hundreds of such cadences have been invented by drill instructors following Duckworth's example (see, for example, Johnson 1983, 1986), and they are a common feature of the boot camp experience in all the services.

By those who employ them, marching chants are also known as "Jody calls," or simply "Jodies"; in vernacular usage, "cadence call" and "Jody call" are synonymous terms. The usage arises from the fact that, from the 1940s on, a certain percentage of such calls have referred to the exploits of a fictional character called Jody, a ne'er-do-well who sexually preys upon soldiers' wives and girlfriends while the soldiers themselves are away from home. Carol Burke collected a typical example from a student enrolled at the Virginia Military Institute in 1988:

> Ain't no use in callin' home.
> Jody's on your telephone.
>
> Ain't no use in lookin' back.
> Jody's got your Cadillac.

DOI: 10.7330/9780874219043.c04

Ain't no use in goin' home.
Jody's got your girl and gone.

Ain't no use in feelin' blue.
Jody's got your sister too. (quoted in Burke 1989, 431)

As the reference to the Cadillac suggests, the Jody character is often seen as stealing (or making use of) not only the absent soldier's woman but his material possessions as well. Like Penelope's suitors in the *Odyssey*, his basic functional role in the military narrative is to serve as a threat to the soldier's home and property.

Like soldiers, men in prison are separated for long periods of time from their loved ones, and thus subject to possible predation by Jody-like characters. It's not surprising, therefore, that the marching chant tradition credited to Willie Duckworth has a parallel in the prison work songs of incarcerated African Americans. Folklorist Bruce Jackson's seminal article (1967) on Jody songs, in fact, mentions the military venue only tangentially: The songs he collected came from prisoners at a maximum-security prison farm in southeast Texas. As he notes in his book on southern blues, a mutual concern of those serving in the Army during wartime and those serving prison sentences is the question of "who is doing what with, and to, the woman one left at home" (1999, 167). Jackson has observed this concern not only in blues lyrics but also in African American toasts dealing with the Jody character, where the protagonist (and toast name) is "Jody the Grinder." In the toasts as in the songs, he says, Jody is a contraction of "Joe the," while "Grinder" functions as "a metaphor for a certain kind of coital movement." One example, a blues performed for Jackson by convict Benny Richardson in March 1966, well illustrates the thematic similarity between the prisoners' songs and the soldiers' calls:

Goin' back home to my old gal Sue . . .
My buddy's wife and his sister, too . . .

Ain't no need of you writin' home . . .
Jody's got your girl and gone . . .

Ain't no need of you feelin' blue . . .
Jody's got your sister, too.

First thing I'll do when I get home . . .
Call my women on the telephone . . .

Yeah, yeah, YEAH, YEAH,
Yeah, yeah, YEAH, YEAH.

Gonna settle down for the rest of my life . . .
Get myself a job and get myself a wife . . .

But there are complications here as well. If the typical Jody call references simply a backdoor man who cuckolds the absent soldier (or prisoner),
the first verse of Richardson's song suggests that the enlisted (or incarcerated) man is himself a sexual predator, the fourth one references his multiple partners ("my women"), and the last one hints at a domestic resolution
that looks beyond the complaint of the usual story line. Such complications
appear in marching chants, too, and I will look more closely at them later in
the essay. For now let me just say that in the "normative" Jody call, soldiers
respond to their drill instructor's cue by lamenting the fact that, while they
are in the service, their women and their goods are at the service of another.

"Lamenting," however, may be too simple a word. While it's clear that,
on the surface, Jody calls could not comfort those singing them, they have
nonetheless been embraced with gusto by generations of soldiers separated
from their women. In addition, as I'll discuss later, recent attempts to clean
up the more obscene versions of the calls—and there are many of these—
have been only marginally, and only publicly, successful. As a military folk
tradition, chanting about the guy who has "got your girl and gone" is as
vibrant today as it ever has been.

Why is that? What is the appeal of these chants to the mostly young
men who share them? Why do they persist despite official bans against them?
What function do Jodies serve in today's military units? Drawing on data
from print and online sources, interviews with Arkansas State University
ROTC students and personnel, and my own experiences in the United
States Marine Corps (1969–1973), I devote the remainder of this chapter
to addressing these questions.

DISCIPLINE, FUTILITY, AND RESENTMENT: "NORMATIVE" JODIES

One obvious answer is a pragmatic one. Jody calls in particular, and
marching chants in general, bring soldiers together as a disciplined force.
This was their original intention, and it is no less relevant today than it

was in Duckworth's time. At 120 beats per minute, or 180 beats per minute during "quick time," marching cadences inspire soldiers to coordinate themselves in motion and, in doing so, instill a sense of esprit de corps. They not only break the monotony and fatigue that troops endure during long marches, but they also inculcate the solidarity and "unit cohesion" that is essential to the efficient functioning of any military force.

In some chants, that solidarity is expressed in a directly martial fashion, as the lyrics describe the fate that will befall the hometown seducer once the absent soldier returns and gets his hands on him. Revenge on the seducer appears in numerous cadence calls as well as in such song lyrics as Jimi Hendrix's "Hey Joe." One frequently cited quick-time example goes like this:

> I'm gonna take a three-day pass,
> Can't wait to get Jody in my grasp.

Here's a more elaborate version, collected by Sandee Johnson from an Army private stationed in Ft. Hood, Texas ("PT" is physical training;).

> Here I am in an old two/five,
> Running PT, staying alive.
> Back home Jody's got my wife,
> Gonna take a Chap and end his life.
>
> Driving trucks the whole day through,
> Watch out Jody, I'm a gunnin' for you.
> So I went home, old Jody's scared,
> Said, "Come on back if you dare."
>
> Jody took an aircraft into the sky,
> Said, "Hey, hey Jody, you're gonna die."
> Took that Chap and brought it on line,
> Jody's death is mighty fine. (Johnson 1983, 15–16)

Such fantasies of revenge are common in the Jody repertoire, but they are not the only sentiment with which soldiers greet their humiliation. Paradoxically, the chanters' solidarity often reflects the recognition not of shared valor or labor or commitment or revenge, but of shared futility. This comes out in chants, for example, that emphasize the physical power as well as sexual aggressiveness of the Jody character. Jody is often a nondescript character, but occasionally his physical strength is noted as evidence that, even if the soldier could confront him, it would be no use. The following verses typify this particular sensibility:

> Jody this and Jody that,
> Jody is a real cool cat.

> If old Jody is six foot tall,
> I won't mess with him at all.

> Might as well hide that frown,
> Jody's beat you hands down.

> Jody, Jody, six feet four,
> Jody's never been whipped before.

> Jody is the one who's mad—
> Basic training ain't that bad. (Johnson 1983, 15)

Such acknowledgment of Jody's strength is relatively rare in the marching chant repertoire, however. A more common index of the futility motif is the "ain't no use" phrase that opens many verses:

> Ain't no use in lookin' down
> Ain't no discharge on the ground

> Ain't no use in lookin' back
> 'Cause Jody's got your Cadillac

> Ain't no use in feelin' blue
> 'Cause Jody's got your lady too

Here, as in numerous other cadences, troops are reminded that they are in limbo and will remain so until they die in combat or return home. Like convicts, they must resign themselves to membership (and when a draft is in effect, to involuntary membership) in a circumscribed and predominantly male world in which they cannot realize their dreams and in which their very lives may be suddenly cut short. While they are serving their time, they come to understand there is "no use" in looking beyond their immediate circumstances. They are all in a situation from which there is no escape. But—here's the important part—they are in that situation together. What Austin Bay says about military slang is true as well of the "no use" Jody versions: They tell the soldier, "Buddy, you ain't in this alone" (2007, 5).

A related reason for the popularity of the chants may be that, in advertising the common soldier's powerlessness, they implicitly attack the institution that keeps him powerless—the military establishment itself. As my fellow Marine Mickey Weems shows in chapter 7 of this volume, Marines' affection for the Corps is often inextricable from resentment toward what they call the Big Green Weenie. Cadence calls may implicitly express that resentment. One of my students, another former Marine, once told me that he saw Jody calls as almost "a way of saying fuck you to the instructors, to let 'em know you're not ready to quit yet." They may allow their users to indirectly attack the imbalances of hierarchy (enlisted personnel versus commissioned officers, convicts versus guards) while taking pleasure in undermining the authority of those under whom they serve. They function, therefore, as a safety valve that only insiders understand. In overtly complaining about what the free-world cuckolder is up to, soldiers are covertly registering resentment at those who keep them unfree and therefore unable to defend themselves against Jody.

Both of these rationales for the popularity of chants—the acknowledgment of futility and the venting of resentment—work well as explanations of the "normative" Jodies: those chants where the villain of the piece is Jody himself and the soldier is the innocent victim of his predatory activity. But not all Jodies fit this neat pattern. To account for the appeal of nonnormative Jodies, we need to bring more nuance into the analysis.

SEX, GUILT, AND PROJECTION

An interesting step in this direction has been provided by political scientist Michael Hanchard (1998) in an essay on the iconicity of the Jody character in the African American community. Like Jackson, Hanchard does not address Jody calls explicitly, but focuses on the character's appearance in toasts, stories, and especially blues lyrics. Nevertheless, his analysis, which stresses the domestic insecurities revealed by the character, is still a helpful key to understanding Jody calls.

In Hanchard's reading, the cuckolder Jody is best understood not as an ill-willed free agent but as one element of a social dynamic whose complications reveal the "yearnings as well as the insecurities and passions of black men and women who are somehow connected to [Jody], or whose relationships are mediated through him" (476). What counts in his analysis is the economic as well as romantic relationships that reflect types of dominance and subordination (who controls whom), and what the "villainous" Jody signifies, in the political economy of the African American community, is how fragilely constructed is black male identity in an environment where men are encouraged to perform both as stable breadwinners and as erotic actors whose recklessness threatens stability. The sad irony is that, in order to provide security for his woman, a man must be absent from the home, and thus (at least temporarily) unable to satisfy his wife's sexual urges. The Jody character, whose social impact is "less individual than it is relational," is a "liminal figure" pointing to that paradox, and he can be understood only within that matrix. "By himself, Jody has no identity" (477).

But it is not just physical absence that is the root of the problem. Drawing on Johnnie Taylor's blues "Jody's Got Your Girl and Gone," Hanchard argues that the man's inability, or unwillingness, to support his woman emotionally—to make her feel wanted—is what really invites Jody in and makes him attractive. Here are the Taylor lyrics:

> Every guy I know, tryin' to get ahead,
> Workin' two jobs till you're almost dead,
> Working your fingers down to the bone,
> Now a cat named Jody sneakin' around in your home
> Now there's a cat named Jody in every town
> Spendin' lots of cash, just ridin' around
> Ride on Jody, Jody ride on with your bad self
>
> [refrain]
>
> Ain't no sense in goin' home
> Jody's got your girl and gone
> When you get home after workin' hard all day
> Jody's got your girl and he's gone away
>
> [refrain]
>
> When you discover that your role is neglected
> It'll be too late to give your woman respect
> You hunt down Jody dead or alive
> Ten thousand dollars for Jody's hide
>
> [refrain]
>
> You'll meet him alone one day
> Jody took my girl and he's gone away

Had enough sense to express himself
Told her how good she looked
Told her how cute she walked
Told her how pretty she talked

[refrain]

The song begins conventionally enough, with the "good" man working hard and "bad" man Jody romancing his woman in his absence. We can hear in this song, too, the "ain't no use" trope so common in marching chants. But in the second verse, the plot thickens, as the good man is seen as having "neglected his role" and failed to give his woman the "respect" she was due. He may comfort himself with the idea that he will some day hunt Jody down and kill him, but that comfort is complicated by his recognition that it was the backdoor man, in his absence, who treated his woman right. It was Jody who "expressed" what the woman wanted to hear: how "good," "cute," and "pretty" she was. When women in blues clubs hear this song, Hanchard says, they often cheer. "Disrupt[ing] the male-identified narrative of betrayal and revenge" (1998, 479), they understand that, if men gave their women the proper emotional support—what Aretha Franklin famously sang as "R-E-S-P-E-C-T"—the enticements of Jody would be less attractive.

Let me bring this discussion back to marching chants by noting that, in some Jodies—a minority, to be sure—it is the soldier himself who plays the spoiler role. It is he who subverts domestic stability either by failing to give his woman the proper attention or by becoming, like Jody himself, a home-wrecking seducer. The first scenario appears in an Airborne chant, where the airman's apparent preference for the military over his home life results in his losing the latter forever:

I got a girl that lives out west
Thought this Airborne life was best
Now she's somebody else's wife
I'll be jumping for the rest of my life. (Johnson 1983, 23)

In the second scenario, the soldier, presumably on leave, is discovered by an irate husband and forced to flee:

I look down the road and what did I see
My girlfriend's husband coming after me

He's got a gun and he's got a knife
Hey, I'll be running for the rest of my life. (21)

In both of these chants, the absent soldier, far from being an innocent victim, is the cause of domestic disturbance. This is not the normative pattern
in Jody calls, but it appears often enough to make us wonder whether or not
the outraged pain implicit in the norm may not function, on a deeper level,
as a transformation of the guilt that some soldiers may feel at their own
infidelities (or imagined infidelities). Could a kind of self-torture be part of
the Jody calls' appeal? A self-torture that functions as punishment for having
done (or thought of doing) harm to one's woman?

Given the nature of military life, the question is not an academic one.
Eric Eliason, who served as a chaplain with US Special Forces in Afghanistan,
tells me that the most common anxiety expressed by soldiers there was fear
that their wives or sweethearts were "messing around." My own military
experience confirms that picture. I know of many instances while I was
stationed at Camp LeJeune, North Carolina, when fellow Marines went
AWOL in order to return home in an attempt to prevent a spouse from continuing an illicit affair. Ironically, after the military police dragged these poor
souls back to base, they often had to serve time in the brig before returning
to their unit, adding more time to their military service and extending their
time away from their perhaps unfaithful loved ones.

So infidelity is a constant concern among deployed troops, one that is
registered not only in the name "Jody" for the soldier's cuckolder, but also in
the name "Suzy Rottencrotch" for the cheating wife or girlfriend (MacNevin
2001; Knight, 2002–2011). However, given that a young male in the field
is every bit as likely to cheat on his sweetheart as she is to cheat on him,
perhaps Jody calls function not just as laments but also as projections, that
is, transfers onto another of one's own sins. Deployed servicemen are not
immune to temptation, especially when, for example, a troop ship reaches a
port where prostitutes are eagerly awaiting fresh sources of revenue. Yielding
to such temptations, a young man may find that he is not so different from
the Jody character about whom he chants.

A Jody call might then offer the cheating male soldier a chance to publicly admit to his own transgressions by projecting them onto his wife and
her illicit lover. Hanchard suggests that Jody "lurks within all of us, male
and female" (1998, 497). Perhaps soldiers' chanting about Jody's sexual
prowess allows them to express, in a veiled fashion, their envy of his success

with women, their guilt at embracing his desires, and a secret glee in being punished for that embrace. His disguised confession, for which the Jody call is a vehicle, includes the punishment of realizing that his wife's love is no longer only his to enjoy.

VARIATION, OBSCENITY, AND CONTINUITY

The appeal of the Jody character is further complicated by the fact that, in some cadence calls, there is little or no evidence of the normative pattern. I'll give two examples from the repertoire of Sergeant First Class Edward Smith, an instructor for Arkansas State University's ROTC program who has had much experience both in teaching cadence calls to cadets and in drilling troops while he was in an artillery unit with the 101st Airborne. When I interviewed him, he read the following call from his desk computer:

> Mama told Jody not to go downtown
> Marine Corps Recruiters are hangin' around
> Jody didn't listen and went anyway
> Now he is livin' the Marine Corps way. (USMC Hangout)

In this chant, Jody, who because of his name might be expected to stay home and seduce soldiers' women, is recruited into the Marine Corps himself. This not only blunts his sexual potential but also—building on the point I've just made—suggests a link between seducer and serviceman that the more traditional chants do not express. The unspoken subtext here might be that Jody is just like us: a sexual adventurer now under the control of the Corps. A second chant that Sergeant Smith offered—one of his own invention— presented an even more interesting variant on the "girl and gone" theme:

> Mama told Jody not to go downtown
> They had too many artillery men hangin' around.
> Jody didn't listen and went anyway
> Now she is laid on her back the artillery way.

Here, taking advantage of the androgynous nature of the name "Jody," Smith subverts the norm even more dramatically. Jody becomes a woman who, oblivious to her mother's warning, hangs out with military men who take sexual advantage of her. One might see a hint of gang rape here, or simply a reflection of the adolescent bluster that is so often an element of boasting about a given unit's prowess. In either case, the feminizing—or emasculation—of the Jody character in this chant may serve as quiet revenge on the normative seducer: a way for potential victims of a "sneaking around" character to turn the tables and make him/her the victim.

Like Willie Duckworth, Sergeant Smith is an African American drill instructor charged with teaching mostly young white males how to march in step and sing about Jody as they do. There may be historical payback in this pattern, as black drill instructors from Duckworth on may have practiced some "signifying" on their charges, requiring them not just to accept but to chant about their inability to protect their women, not to mention their own honor, from Jody's advances. It's interesting to note that, at the very least, Jody lore is an African American invention, and its use in the military has long been well established as the province of middle-aged sergeants whipping adolescents into shape.

In doing so, professionals like Sergeant Smith have traditionally relied not only on established chants, but on their own ingenuity in adapting them and in coming up with others. Like other oral folk genres, cadence calls are a dynamic medium, constantly evolving variants in response to social contexts and individual creativity. Smith's example of the female Jody "laid on her back" is but one example of that creativity. One of the reasons for the chants' appeal, I suggest, is precisely their fluidity and openness to lyrical change, even as their rhythmic structure remains constant.

The dynamic and resilient nature of the Jody call tradition can be seen particularly clearly when we consider obscenity. Not surprisingly in an environment that until recently was all male, obscenity has always been a central feature of marching chants, whether or not the Jody character has been in play. When I was in boot camp in 1969, for example, one of our drill instructors marched us to a chant that managed to be sexist and culturally insensitive at the same time:

> I don't know but I've been told
> Eskimo pussy is mighty cold
> Mmm, good! Feels good!
> Is good! Real good!
> Tastes good! *Mighty* good!
> Good for you! Good for me!

This call is now available online (Allison 2006), but that I can remember it more than forty years later attests to its emotional charge. It was an effective vehicle for creating solidarity among young men who had been abruptly removed from the comforts of home and possible romance. Marginally less vulgar but nonetheless sexually explicit lyrics have appeared in hundreds of Jody calls over the years.

So common is obscenity in this genre, in fact, that collectors have routinely felt obliged to expurgate their data and to offer conventional apologies in the service of propriety. In introducing the first of her two volumes of cadence calls, for example, Sandee Johnson observes, "A few Jodies have only five or six words remaining after the vulgarities are edited—thus these verses were omitted." In addition, she continues, "Several of those included may warrant criticism from women due to their sexual overtones" (1983, 1). Such editorial timidity is of course anathema to folklorists, who subscribe generally to Horace Beck's counsel that the first rule of collection is to "take it as it comes" (1962, 195). But it is far from a rare approach among nonfolklorists.

For different reasons, a concern for propriety has also worked its way into military command circles. Since the entry of women into the US armed forces, and since the Tailhook and other scandals have made military leaders sensitive to charges of sexual discrimination, there has been an official ban on chants of a sexually explicit or misogynistic tenor. I agree with Carol Burke that the ban has always had less to do with sensitivity to female recruits' feelings than the brass's concern about negative PR (1989, 438). But in any event, my research suggests that the ban has had less effect than official channels might like to suppose on the actual chanting practices of recruits and other enlistees. The data I have received from veterans attest to the power of obscene language in cadence calls, in particular with respect to Jody calls, and it suggests that, rather than disappearing entirely, obscene calls have simply gone underground. This, for example, is how the situation is described by an Arkansas State student and Army veteran who was enrolled in my American Folklore course after being wounded in Iraq in 2006: "While I was in basic training (BT) and advanced individual training (AIT) in 2002, during the majority of our marches to and from training, the cadences were clean and nonderogatory. However, if we were off the beaten path, where no officer or female could hear you, my drill sergeants would occasionally loosen up and recall an old Jody cadence from their Vietnam-era days. On these rare occasions these cadences actually seemed to boost the overall morale of the troops even more than just the regular bland ones."

What's evident here, beyond the fact that obscene calls survive out of earshot of women and officers, is the fact, as I've noted earlier, that Jody calls—even, or maybe particularly, the obscene ones—bring young men together in difficult situations. Their resilience in the face of official bans attests to their power as solidarity builders, and to their ability to morph, as it were, according to circumstances. My student speaks to that aspect as well, when he describes how the Jody mythology remains strong even in the absence of its traditional marching chant format:

> During my first tour in Iraq in 2006–2007 these cadences were not used in the traditional way anymore. They were more referred to in a joking way, without the song/cadence. For example: when I was wounded during my first tour the guys in my squad gave me hell about how my girlfriend was going to leave me for Jody since I was a "broke dick" [hurt]. So you still have the same types of jokes that originated from these cadences, i.e., Jody taking your women, being a "sick-call ranger." They're just mostly done more discreetly and without the rhythm.

As this student and other informants demonstrate, Jody calls continue in settings in which women—or civilian visitors to a military base—are absent. The publicity of the calls may be restricted, but their popularity among soldiers is alive and well.

RETURN TO NORMALCY

In the 1949 MGM film *Battleground*, following tremendous losses after a squad of the 101st Airborne Division is trapped in Bastogne during the Battle of the Bulge, a dirty and frightened foot soldier, Private Holley, asks his sergeant, "Whatever happened to Jody?" Sergeant Kinnie, played by James Whitmore, responds by enlisting the battered men of I Company in a traditional Duckworth chant. They are marching from the battlefield, passing their replacements:

SERGEANT KINNIE: All right, come on! Come on! What do you want these guys to think, you're a bunch of WACs? Alright, alright, pick it up now. Hut, two, three, four. Hut, two, three, four. Hut, two, three, four. You had a good home but you left . . .

I COMPANY: You're right!

SERGEANT KINNIE: Jody was there when you left . . .

I COMPANY: You're right!

SERGEANT KINNIE: Your baby was there when you left . . .

I COMPANY: You're right!

SERGEANT KINNIE: Sound off!

I COMPANY: One, two.

SERGEANT KINNIE: Sound off!

I COMPANY: Three, four.

SERGEANT KINNIE: Cadence count.

I COMPANY: One, two, three, four. One, two, three, four!

SERGEANT KINNIE: Your baby was lonely as could be . . .

I COMPANY: Until Jody provided company!

SERGEANT KINNIE: Ain't it great to have a pal . . .

I COMPANY: Who works so hard to keep up morale!

SERGEANT KINNIE: You ain't got nothing to worry about . . .

I COMPANY: He'll keep her happy till I get out!

SERGEANT KINNIE: You won't get out until the end of the war . . .

I COMPANY: In nineteen hundred and seventy-four!

SERGEANT KINNIE: Sound off!

I COMPANY: One, two.

SERGEANT KINNIE: Sound off!

I COMPANY: Three, four . . .

SERGEANT KINNIE: Sound off!

I COMPANY: One, two, three, four. One, two, three, four.

It's an extraordinary moment. On the front near Bastogne, the soldiers are low on fuel, rations, and ammunition, and they are stuck in a pervasive fog that makes movement and identification difficult and prevents their relief by Allied air support. In this difficult if not desperate situation, a soldier asks for news of the proverbial seducer, and Sergeant Kinnie, smiling, is happy to comply. He is happy to take advantage of the opportunity to return his beleaguered charges to a familiar mindset: Things are terrible here but eventually they will return to normal and yes, Jody is indeed back home with your wife.

That weird type of consolation provides as much normalcy as the scared soldiers can hope for given their circumstances. However disagreeable the idea of Jody may be, at least thinking of him means that, for a few moments at least, they don't have to think of the battlefield. My Arkansas State student who was wounded in Iraq spoke to this consolatory aspect of the Jody material when he observed that Jody jokes and cadences remained meaningful to soldiers because they helped to mute anxiety with humor. "It's easier to turn a nervous thought and/or situation into something funny," he told me, "than worry about things that are out of your control."

For soldiers like Private Holley, thinking about Jody is accompanied by marching—another element of the familiar and the predictable and one that, by disciplining the soldiers' bodies, also helps to keep their minds off their plight. But of course marching is also a sign of the inescapable fact that these soldiers are not going home anytime soon. As long as they are in uniform, they must continue to be alert to the potential horror of death on the battlefield from some stray bullet, bomb, or grenade: The ordering of attention provided by synchronous movement is always partial, always threatened by the chaos that accompanies what the Prussian military theorist Carl von Clausewitz famously called the "fog of war." For the young men of I Company, that metaphorical fog is potentially more unsettling than the physical fog that traps them at Bastogne. It means that, despite Kinnie's good intentions, they are still subject to the unplanned, the out-of-step, the randomly lethal. In a firefight or other combat situation, virtually anything can happen, and often does. Synchronous movement may disguise that possibility, but it cannot obliterate it. Hence the paradoxical attraction, to Holley and his comrades, of the Jody chants. They turn the mind to a familiar source of anxiety and away from the alien source that is the battlefield reality.

For soldiers facing death, moreover, there may even be a certain level of pleasure in contemplating the exploits of this ambivalent character. On the one hand, Jody's deviousness while enjoying a wayward spouse's love can only bring pain to her infantryman husband. On the other hand, though, the soldier may vicariously partake in Jody's sexual pleasures—ones that he himself can only imagine. Jody chants, I submit, may give soldiers access to a voyeuristic sexual fantasy that embodies both jealous rage and imagined marital bliss. The character's ability to exact such diverse emotions may help to explain why soldiers willingly enter his world. Whether or not that world is make-believe, it has an inherent appeal as a reminder of conjugal pleasure, and it is a world at the very least that is far from the battlefield.

Compared to this distant and admittedly ambiguous fantasy world, the real world of the battleground soldier is uncharted territory. Being "in country," even when the bullets are not flying, means by definition being away from home, and therefore in a place, as Defense Secretary Donald Rumsfeld once put it, of "unknown unknowns." The sense of dislocation associated with that condition is registered in military slang for some deployments. Armed forces personnel assigned to the Middle East, for example, may say that they are in Iraq or Afghanistan, but they are also likely to say that they are "in the sand," suggesting an unspecified tract of unmappable real estate. Dislocation is suggested also by the many Jody calls that begin "I don't know but I've been told." Soldiers who chant that line acknowledge symbolically that they are about to enter unknown territory and engage in activities about which they know little, with even that little based only on hearsay.

Even when they enter that unknown territory, much about military life remains unfamiliar. Traditional complaints about military food, clothing, gear, and equipment, therefore, might be seen not as soldiers' "objective" assessments of inferior materials, but as signs that everything here, in country, is different from home. Jody calls may offset that daily recognition. Even as they remind the soldier that he is not there, they evoke fond memories of a civilian realm to which, if he is lucky, he may one day return. Until (and if) that time comes, the soldier is better off to resign himself to his fate rather than worry about his girl's unfaithfulness. Perhaps there truly "ain't no use in goin' home," especially if Jody's got your girl and gone. The sense of futility evident in so many Jody calls may simply force the soldier to think of more immediate concerns, like survival on the battlefield, rather than a return to domestic normalcy.

"IF I DIE"

But every combat soldier knows, of course, that there is a possibility he will not return home. That possibility is registered in those Jody calls in which the soldier comes to terms with the likelihood of dying in battle. Carol Burke heard a good example of this common variant while she was teaching at the US Naval Academy in 1987. ("Leaning rest" is the beginning position for push-ups; a "K-bar" is the standard combat knife.)

If I die in a combat zone,
Box me up and send me home.

Tell my Momma I did my best
Then bury me in the leanin' rest.

Mothers of America now don't you cry.
Marine Corps way is to do or die.

Place a K-bar in my hand
And I'll fight my way to the Promised Land! (1989, 432)

The "If I die" premise, which is a common trope in Jody calls, is often resolved by the victim being buried with the sexual comfort that his military service has denied him. This example, found in Johnson's collection, dates from World War II:

> If I die on the German line
> Bury me with a sweet Fraulein.
>
> Pin my medal on my chest
> Tell my girl I did my best. (1983, 27)

Presumably, the soldier's girl is not the "sweet Fraulein" to whom he refers. A cold war variant, not appearing in Johnson's collection but included in a study by Kent Lineberry, contains the following verse:

> If I die on the Russian front
> Bury me in a Russian cunt! (2003, n.p.)

Lineberry also cites a long Jody call in which an Airborne soldier, about to parachute into hostile territory, reflects on both of these common marching chant tropes: the likelihood of his death and the consolation he might receive from being buried with a woman. (A C-130 is a transport plane; "my man" is slang for parachute; PLF stands for "parachute landing fall.")

> C-130 rolling down the strip
> Airborne daddy on a one way trip
> Mission uncertain, destination unknown
> We don't know if we'll ever come home
> Stand up, hook up, shuffle to the door
> Jump right out and count to four
> If my man don't open wide
> I got another one by my side.
> If that one should fail me too
> Look out ground I'm coming through
> Slip to the right and slip to the left
> Slip on down, do a PLF
> Hit the drop zone with my feet apart
> Legs in my stomach and feet in my heart
> If I die on the old drop zone
> Box me up and ship me home

Pin my wings upon my chest
Bury me in the leaning rest
If I die in the Spanish Moors
Bury me deep with a case of Coors
If I die in Korean mud
Bury me deep with a case of Bud
If I die in a firefight
Bury me deep with a case of Lite
If I die in a German Blitz
Bury me deep with a case of Schlitz
If I die don't bring me back
Bury me with a case of Jack.

In this elaborate celebration of America's military actions abroad, the soldier is no longer concerned with what Jody is doing back home, but only with the easy comfort of beer or whisky ("Jack" is Jack Daniels bourbon). It is a different comfort from that of sleeping with a woman, but it shares with the comfort of sex an appreciation of physical pleasure that the strictures of military life often make inaccessible.

What many of the "If I die" verses suggest, then, is a kind of last-ditch, carpe diem sensuality. In the face of death, the soldier thinks only of accessing the two things he deems worthy to pursue: sex and alcohol, in pretty much that order. But of course, such hedonism is exactly what the Jody character represents, what he brings to an established relationship as a disruptive force. In a sense, then, the threat of death turns the soldier away from duty, away from marital fidelity, and toward the same wantonness for which Jody is condemned. In the "if I die" chants, which express a sentiment common to many Jody calls, the soldier becomes a kind of foxhole existentialist, seizing the day of pleasure as he prepares to die. In this narrow, bare-bones focus on immediate gratification, he joins hands emotionally with his at-home antagonist, becoming a Jody-like character, wanton and free.

Perhaps to young men who are anything but free, this reveals the deepest appeal of the Jody character. He is what they, as contract-bound soldiers, fantasize about being, if they could only get out of uniform and into someone's bed. Like other "good bad men" of American legend, such as outlaws and gangsters, the Jody character survives not in spite of embodying a threat to disciplined domesticity, but precisely because he does. For young men away from home—many of them for the first time—that threat may be as secretly appealing as it is fearful.

CONCLUSION

In his 1967 essay on prison songs, Bruce Jackson echoed Private Holley's query to his sergeant by posing the question "What happened to Jody?" In this essay, building on my former studies of military lore (2003, 2006), I have in effect raised that question for a third time. I hope I have provided some answers that go beyond the conventional view of Jody as, in the words of one of my student veterans, simply "the guy who fucks your old lady while you're gone." I've tried to show here that while the normative Jody does fit that stereotype, contemporary servicemen respond psychologically to this character with sensibilities considerably more complicated than the simple urge to indulge in humiliating fantasies. The Jody of a thousand marching chants can be victimizer or victim, male or female, civilian or soldier, and the attitude of soldiers who chant about him while doing quick-step involves a wide array of psychological elements, ranging from hatred, suspicion, and fear on the one end to envy, guilt, and half-suppressed desire on the other. Soldiers' reactions to Jody calls are probably as varied and complex as the calls themselves.

Their complexity, moreover—what folklorists would identify as their textual variation—probably has a great deal to do with their enduring appeal. They have been a feature of military culture now for three generations. In an era of women in the ranks and of institutional sensitivity to public appearance, the most vulgar examples may have gone underground. But Jody calls survive in oral culture, out of sensitive earshot, fulfilling psychological needs for those in harm's way. They may not be as visible as they were in Willie Duckworth's day. But "in the sand" and in the barracks, they are far from dead. Like the backdoor man Jody himself, they may only be hiding.

WORKS CITED

Allison. 2006. "Eskimo Song." Posted 16 September on Urban Dictionary. http://www.urbandictionary.com/define.php?term=eskimo%20pussy.

Bay, Austin. 2007. "Embrace the Suck." In *A Pocket Guide to Milspeak*. New York: The New Pampleteer.

Beck, Horace P. 1962. "Say Something Dirty!" *Journal of American Folklore* 75 (297): 195–9. http://dx.doi.org/10.2307/537721.

Burke, Carol. 1989. "Marching to Vietnam." *Journal of American Folklore* 102 (406): 424–41. http://dx.doi.org/10.2307/541782.

Burns, Richard. 2003. "'This Is My Rifle, This Is My Gun . . .': Gunlore in the Military." *New Directions in Folklore* 7.

Burns, Richard. 2006. "'I Got My Duffel Bag Packed/And I'm Goin' to Iraq': Marching Chants in the Military." *Ballads and Songs—International Studies* (BASIS) 4: 166–75.

Hanchard, Michael. 1998. "Jody." Critical Inquiry 24 (2): 473–97. http://dx.doi.org/10.1086/448881.

Jackson, Bruce. 1967. "What Happened to Jody." *Journal of American Folklore* 80 (318): 387–96. http://dx.doi.org/10.2307/537417.

Jackson, Bruce. 1999. *Wake up Dead Man: Hard Labor and Southern Blues.* Athens: University of Georgia Press.

Johnson, Sandee Shaffer. 1983. *Cadences: The Jody Call Book, No. 1.* Canton, OH: Daring Books.

Johnson, Sandee Shaffer. 1986. *Cadences: The Jody Call Book, No. 2.* Canton, OH: Daring Books.

Knight, Glen B. 2002–2011. "Suzy Rottencrotch." *Unofficial Unabridged Dictionary for Marines.* http://oldcorps.org/usmc/dictionary/s.html.

Lentz, Bernard. 1955. *The Cadence System of Teaching Close Order Drill and Exhibition Drill.* Harrisburg, PA: Military Service Publishing Company.

Lineberry, Kent. 2003. "Cadence Calls: Military Folklore in Motion." *Missouri Folklore Society Journal* 25.

MacNevin, Suzanne, comp. 2001. "Military Slang." http://www.lilithgallery.com/articles/suzyrottencrotch.html.

USMC Hangout. "Marine Corps Cadences." http://www.usmchangout.com/usmc/facts/cadences.htm.

5

Upper Echelons and Boots on the Ground
The Case for Diglossia in the Military

Elinor Levy

Seeing as how the VP is such a VIP shouldn't we keep the PC on the QT because if it leaks to the VC he could end up an MIA and we could all be put on KP.

—Robin Williams as Adrian Cronauer in *Good Morning, Vietnam* (1987)

When young men and women make the transition from civilian to military life, they are subjected to what George Rich and David Jacobs once described as "a process of forced acculturation" to make them culturally competent in their new situation (1973, 164). To acquire cultural competence, they must learn and internalize a wide array of new rules, procedures, and traditions—all the military-specific matters that in this volume we are calling "warrior ways." At the same time, they must learn a new form of communication: a new language, in fact, which at first must seem as incomprehensible to them as any foreign tongue. This new language, which is based on everyday English but often doesn't sound much like it, is a principal means through which acculturation takes hold.

That military language serves, no less than bugle calls or forced marches, to create solidarity among recruits is, I think, an inarguable proposition. The term *common language*, however, can be misleading. In exploring the "common language" that is usually identified as "military speech," we should be careful to distinguish between two different forms of this specialized jargon. On the one hand, there is the acronym-heavy and (to outsiders) barely intelligible language that is promoted by the Department of Defense and other government bodies. This is the "peculiar artificial language" (Elting, Cragg,

DOI: 10.7330/9780874219043.c05   99

and Deal 1984, 227) that identifies a soldier's automobile as a POV (privately owned vehicle) and that calls the unintentional killing of civilians "collateral damage." On the other hand, there is the unsanctioned, sometimes vulgar, and often transgressive slang created by soldiers and other service members themselves—the speech that gave us the mocking acronym SNAFU (for "Situation normal: all fucked up") and the insulting epithet "shake and bake" for a newly minted officer (Burke 2003). While these two forms of military speech have different origins, different emotional tones, and different appropriate venues, in order to function effectively in today's American armed forces, one must be linguistically competent in both forms. To "talk the talk," in today's military, then, requires something more than acquiring a common language. It requires becoming equally familiar with two very different manifestations of the English language—what sociolinguists would call two different registers. It requires becoming, in a sense, sociolinguistically bilingual. In this essay, I explore the implications of that fact.

Since the first type of military speech is endorsed by, indeed produced by, the Department of Defense, it has sometimes been called "Pentagonese." But because it is in general use far beyond the reaches of the Washington bureaucracy, I will refer to it with the more expansive term *officialese*. For the second type of speech—the slang of the enlisted personnel with boots on the ground—I have coined the term *enlistic*. I will be using these two terms throughout the essay, and I will be referring to the languages they describe as specialized "varieties" of military English. I will also make it clear later in the essay that the two varieties, rather than being mutually exclusive "sociolects," are more like different registers of speech that constantly interpenetrate and reinforce one another.

I borrow the term *variety* from Charles Ferguson, the American linguist whose 1959 essay "Diglossia" remains a seminal source for our understanding of bilingualism, and particularly of that form of bilingualism in which a single language "divides," as it were, into a more formal and less formal variety—as, for example, in cases where a standard language (like the "Queen's English") exists alongside a less prestigious regional dialect (like London Cockney). Ferguson referred to the "standard" dialect in cases like this as the High, or H, variety, and to the regional dialect as the Low, or L, variety. In this essay, while admitting that officialese and enlistic are not, strictly speaking, English dialects, I will suggest that Ferguson's H-L distinction can still help us illuminate these two varieties of military speech. While acknowledging the ways in which my "O-E" distinction departs from Ferguson's model,

I will try to see how useful a concept diglossia might be in explaining why military folks speak as they do.

DIGLOSSIA: THE FERGUSON MODEL

Ferguson's essay delved deeply into the phonological, grammatical, and lexical complexities of diglossia as it existed in four "defining" languages: Arabic, Greek, German, and Haitian Creole. As fascinating as his analysis is, much of it is irrelevant to a discussion of officialese and enlistic, and I will therefore focus on those of his observations that can help us illuminate the O-E division in military speech. Four of these are paramount.

First, Ferguson defined diglossia as a "particular kind of standardization where two varieties of a language exist side by side throughout the community, with each having a definite role to play" (1980 [1959], 232). What's critical here is the notion of concurrent usage. It's not that one variety is used for a while and then replaced with another, but that both are used simultaneously, and that the double usage has become stabilized—in Ferguson's wording, standardized. The proponents of the H form may condemn the L form as substandard, but they do not—as long as diglossia continues to exist—succeed in driving it out of usage. Nor do some speakers use only one language and not the other. In a truly diglossic speech community, everybody is familiar with both varieties.

Second, each variety's "definite role to play" is directly connected to its prestige value. The L variety is the language of the home and the street: In fact, it is most people's "native" variety, the one they learned as children. The H variety has to be picked up later, in formal education. It is what Ferguson calls a "superposed variety," and it carries more prestige than the childhood version, even among those who routinely speak the L variety. In some situations, in fact, the privileging of the standard is so extreme that "H alone is regarded as real and L is reported 'not to exist' . . . There is usually a belief that H is somehow more beautiful, more logical, better able to express important thoughts" (1980 [1959], 237).

Third, while much of the vocabulary of the High and Low varieties is shared, there is also a sizable body of "paired items," calling for totally different terms for the same thing depending on whether one is speaking the H or the L variety. Even when "the range of meaning of the two items is roughly the same," the use of one rather than the other "immediately stamps the utterance or written sequence as H or L" (1980 [1959], 242). Ferguson gives an example from Greek, where the H word for wine is *inos* and the L word is *krasi*. "The menu will have *inos* written on it, but the diner will ask the waiter for *krasi*" (243). At the lexical level, in short—at least for certain words—the border between H and L is strictly patrolled.

Finally, and most important, there is a rigid "specialization of function" for the two varieties. The appropriate cultural use of each speech repertoire is a key to understanding diglossic communities. In general, the High or standard language is used for "official" purposes such as religious, governmental, and educational exchanges; it is also used in classical or otherwise prestigious poetry, and in fact is generally thought of—even by Low speakers—as the language of great literature. The Low language is used for conversations with family and friends, for addressing children and servants, and in folk literature. The two realms, according to the cultural norms, can overlap "only very slightly." "The importance of using the right variety in the right situation can hardly be overestimated. An outsider who learns to speak fluent, accurate L and then uses it in a formal speech is an object of ridicule. A member of the speech community who uses H in a purely conversational situation or in an informal activity like shopping is equally an object of ridicule" (Ferguson 1980 [1959], 236).

In the diglossic communities described by Ferguson, then, the two varieties of language are distinct not only in terms of vocabulary but also in terms of social appropriateness. "Mixing" the varieties is not allowed, as using the wrong form of the language will be construed as incompetence and could even be seen as alienating or insulting. Even though someone in a Creole-speaking community may know the two varieties well, he or she will be careful at home to refer to people as *moun* and to use the French word *gens* when out in public. Knowing when to switch from one variety to the other is central to a diglossic community's cultural competence.

With that sketch of "natural" diglossia in mind, let me turn now to the "peculiarly artificial" realm of military speech. By looking first at officialese and then at enlistic, I will try to determine to what degree "milspeak" (Bay 2007a) can be considered diglossic, and to what degree it departs from Ferguson's model.

OFFICIALESE: THE H VARIETY?

In that model, H is typically a written variety of the common language that carries higher prestige than L because it is associated with literature, religion, or political authority. Literature and religion rarely enter into the picture with regard to military speech, but political authority does, and in this respect at least officialese, which I am calling the O variety, does seem to function like Ferguson's H. As the term suggests, officialese is the language of military officialdom and formal communication, and its importance as the formal variety is so well understood that there is even an official "bible" of its usage: the Department of Defense (DOD) *Dictionary of Military and Associated Terms*. A random sampling of terms from this massive compilation will give the flavor of milspeak in its O persona.

According to the DOD official lexicon, a "reportable incident" is "any suspected or alleged violation of Department of Defense policy or of other related orders, policies, procedures or applicable law, for which there is credible information." "Fire support coordination" is "the planning and executing of fire so that targets are adequately covered by a suitable weapon or group of weapons." "Reachback" is "the process of obtaining products, services, and applications, or forces, or equipment, or material from organizations that are not forward deployed." A "joint captured materiel exploitation center," or JCMEC, is "a physical location for deriving intelligence information from captured enemy materiel." An argument can certainly be made that such bulky phrases, like the legal phrases they resemble, have the advantage of precision on their side. But this precision is generally purchased at the expense of economy, making officialese the polar opposite of the plain style that George Orwell advocated in his famous essay "Politics and the English Language" (1984 [1946]). It employs multiple qualifiers, strings of prepositional phrases, and clusters of often abstract terms to convey messages that may in general be precise but are rarely verbally elegant or colorful.

The number of terms that the Department of Defense believes needs such treatment runs, in the current edition of the dictionary, into the hundreds. That's not counting the many officialese terms that appear chiefly as acronyms. From their first days in the service, young officers in training learn that they are enrolled not in Officer Candidates' School or the Reserve Officers' Training Corps, but in OCS or ROTC (often pronounced "rot-sea" by outsiders). They are tested there on their familiarity with other acronyms, and by the time they (or their enlisted counterparts) leave the service, they have become culturally competent in using abbreviations that to civilians look like alphabet soup. Indeed, fluency in "acronymese" is a distinguishing characteristic of military life, and a substantial portion of the O lexicon is devoted to such abbreviations.

Some of them, in calling to mind the longer terms that they signify, seem to fulfill their ostensible purpose of streamlining communication. The terms XO for "executive officer" and SECNAV for "secretary of the Navy," for example, are helpful in this way. So too are the terms POV for "personally owned vehicle" and MRE for "meals ready to eat." Others make a kind of comic sense: consider WISDIM for "Warfighting and Intelligence Systems Dictionary for Information Management," FLIP for "flight information publication," or the whimsical IMP for "information management plan." But in many cases, the distance between acronym and meaning is too great to suggest efficiency, or even comprehensibility. "Write once, read many" may be sage advice, but is it really called to mind by the acronym WORM? We may reasonably question whether communicative efficiency is achieved by such terms as FEZ ("fighter engagement zone"), FEBA ("forward edge of the battle area"), and JRX ("joint readiness exercise").

The longest military acronym I've uncovered is ADCOMSUBORD-COMPHIBSPAC. Clocking in at twenty-two letters, it stands for "Administrative Command, Amphibious Forces, Pacific Fleet Subordinate Command" (Towell and Sheppard 1985, 47). That's precise, all right, but so laughably bulky (and out-of-date) that it's no longer listed in the DOD *Joint Acronyms and Abbreviations Dictionary*. Encodings like this one make us understand why critics of military jargon, following Orwell, argue that its real function is not elucidation but rather the obfuscation of meaning. Pentagonese, for example, has been denounced as "a peculiar artificial language developed by the aborigines of the Puzzle Palace for the express purpose of confusing the troops" (Elting, Cragg, and Deal 1984, 227–228). "Express purpose" here is probably tongue-in-cheek, but many critics have agreed that "confusion," from the military's perspective, is not an unwelcome effect of its jargon, especially when it is evading responsibility for unseemly actions.

Gregory Clark, in a comprehensive dictionary of terms from the Vietnam War, gives a good example. A common English description of an attack on a village might be "We napalmed some enemy troops." An officialese version would be the less aggressive sounding: "Multiple dispersed soft targets, in the vicinity of specified coordinates were engaged with soft ordnance." Similarly, "Some collateral damage was unavoidable" sounds more humane than "We accidently bombed a hospital." "A reachback operation was implemented" sounds more efficient than "We went back to get what we forgot." Perhaps MILDEC sounds more honorable than "military deception." And surely "The troops were ordered to break contact and make a retrograde movement from a numerically superior enemy force" sounds less humiliating than "The outnumbered troops were told to retreat" (Clark 1990, 325).

Based on these examples, can officialese be considered the H variety of military language? The answer is mixed. As the product of official circles, it is certainly a jargon of formality and command, emanating from and approved by the institution's upper echelons. It is certainly a specialized, artificial lingo, one that must be studied and memorized by adults rather than picked up naturally during childhood. And there is even a certain quasi-mystical air to some of its lexicon, a mysterious differentness from everyday English that is presumably meant to enhance its aura of special-ness. In all of these senses, it does resemble Ferguson's notion of a High language variety.

But there's one big difference. The use of officialese is not restricted to special venues, or to special people: Everyone, from general to grunt, uses it all the time—indeed, is expected to use it all the time. This means that, compared to a traditional H variety, O is more democratic and less prestigious. It is also, arguably, less "beautiful" than a true H variety. There are some elegant exceptions to this rule, but in general it's fair to say that no one would accuse officialese of being "poetic." It is a formal, somewhat antiseptic mode of speech that is "high" in the sense that it comes from on high, but so universal in adoption that it is as likely to be heard in the bar-racks as in the officers' club. I will say more about this universal availability of O later in the chapter.

ENLISTIC: THE L VARIETY?

If officialese is the language of management and command, then enlis-tic is the working soldier's answer: a variety of milspeak that is as informal, flexible, and crude as officialese is formal, rigid, and euphemistic. If offi-cialese is "top down," then enlistic is decidedly "bottom up." If officialese is a constructed language, carefully crafted and vetted by those in author-ity, enlistic is a natural expression of a "boots on the ground" sensibility—unpredictable, creative, and often transgressive. It is also, as Joseph Roulier wrote of GI slang just after World War II, very much a response to military strictures. "As though it were a direct reaction against the depressing olive-drab motif of his surroundings, the soldier's talk abounds in colors of the brightest hues" (1948, 17). In this section I'll look at some examples of that abounding color, and then consider the legitimacy of seeing E as "Low" military speech.

As a "direct reaction" to military drabness, enlistic is not only colorful but highly emotional as well, and its emotionality often reflects a personal dissatisfaction with one or another aspect of military life. Thus while officialese is dispassionately descriptive, enlistic can be passionately, and often satirically, critical. The love-hate relationship that enlistees have with the service is of course well known: Examples of this abound in popular culture, and in several of the essays in this volume, especially those by Weems (chapter 7), Gilman (chapter 9), and Gillespie (chapter 6). Enlistic is one way in which that ambivalent attitude is expressed. It allows a soldier without significant authority to project an "image of himself and the image of himself he wishes to portray to others" as well as "his attitudes toward the authoritative situation" (Elkin 1946, 417). Given the nature of military assignments, it's not surprising that those attitudes aren't always cheerful. While officialese may require soldiers to maintain the traditional stiff upper lip, enlistic gives them free rein to express frustration, anger, fear, and boredom as well as the black humor that is common in professions involving violence and death (similar types of humor can be found among firefighters, police, and health care workers). In an uncertain and frequently dangerous arena, enlistic generates terms that are "conveniently expressive" (Roulier 1948, 16).

Violence is often at the center of that expressivity. Tank driver slang for infantry, for example, is "crunchies"—a darkly humorous evocation of the sound made by a tank rolling over a body. "Shake and bake," which in boot camp slang means a newly commissioned officer, came to mean in Vietnam an attack sequence of high-explosive bombs (the shake) followed by napalm (the bake)—a grimly humorous "euphemism" that actually tells the brutal truth (Bay 2007a, 42). In the Middle East today, Iraqi fighters' practice of spraying an area with indiscriminate fire is known as the "spray and pray" technique and as the "death blossom" (globalsecurity. org; see also Eliason's chapter 3 in this volume). And enlistic often evokes a unit's battle zone vulnerability. The Logistics Support Area Anaconda, a major supply base near Balad, Iraq, is known by those stationed there as "Camp Bombaconda." The slang term for Balad itself is "Mortaritaville" (Bay 2007b; Glowka et al. 2008, 86). In hundreds of similar usages, enlistic tells the grim truth about warriors' situations that the "clean" language of officialese tends to disguise.

If violence is central to enlistic discourse, so too is sexuality and a concern for personal virility as a measure of worth. In the largely male world of the military, it is not surprising that vernacular speech should contain a large number of sexual, sexist, and sexually derogatory terms. The enlistic vocabulary is rich in this regard. S. G. Kenagy noted this when he described "the emphasis of the language in projecting the image of virility, its concern with the control of power, its expressions of anxiety and aggression, and its reflection of sexual fear and ambivalence" (1978, 90) He was speaking in particular about airmen's jargon, where an ace flyer is called a "hot stick" and pilot's wings are referred to as "leg spreaders" from their supposed ability to attract and bed impressionable women (93). But the point could apply just as well to enlistic more generally. Its vocabulary has changed a good deal over the years, but it has never been short of terms that refer to the genitals, to sexual conquest, or to various forms of "fucking," many of them more aggressive than amorous.

An isolated outpost, for example, may be referred to as "Bum Fuck Egypt" or "Big Fucking Empty." A chaotic situation can be a "gaggle-fuck" or a "clusterfuck." "PFM" is Air Force slang for "pure fucking magic," offered as an explanation of technical matters. "Pump and dump" is to have sex, a "fuck trophy" is a child and, in a humorous adaptation of the NATO phonetic alphabet, "Sierra Tango Foxtrot Uniform," or STFU, is the E slang euphemism for "Shut the fuck up." These few examples are representative of dozens of others in which the young male's favorite Anglo-Saxonism provides "color" to his speech.

Enlistic also employs sexual terms to degrade those not proving to be masculine enough, strong enough, or aggressive enough. Drill instructors in particular are fluent in the use of sexual insults such as "dick," "dickhead," "fuckstick," and "pussy," and it is these middle-aged veterans who, despite official disapproval, generally introduce recruits to the acceptability of vulgarity. In this sense enlistic functions almost as a "natural" language, in which an older parent figure instructs his "children" in the proper use of the oral, rather than official, language.

The enlistic jargon is long on insults, then, but not all of them are sexually tinged. Like soldiers everywhere, speakers of American enlistic have a wide repertoire of terms to describe their enemies, and many of these are as colorful as they are ethnically insensitive. It is doubtful that a press-conscious Pentagon would approve of such terms today, but they are endemic throughout the services, among officers as well as enlisted personnel.

During World War I American doughboys, like civilians generally, spoke of "Jerry" and "Fritz" and "the Hun." In World War II the unofficial but universal term for Germans was "Krauts" (a reference to sauerkraut), while those in service of the emperor were "slants" or "Japs." Vietnam popularized the term "Charlie" for the Vietcong and "gooks" for the Vietnamese more generally—the latter had originally referred to North Koreans (Clark 1990, 204). Today, those serving in the Middle East may in the presence of an officer refer to Afghanis or Iraqis; away from that representative of officialese, they may well use the pejorative term "raghead," or refer to every local as a "hadji," even though that term applies strictly (and honorifically) only to Muslims who have made the hadj, or pilgrimage to Mecca.

But soldiers' animus toward those who are different from themselves isn't always directed at such obvious targets. Enlistic also contains many derogatory terms referring to members of a military branch different from one's own, and even to members of a different MOS (military occupational specialty) within the same branch. I call this subset of the E variety *branch enlistic*. It includes, for example, the term "gravel crushers," a cavalry nickname for the infantry; "the knee-deep Navy" and "puddle pirates," Navy slang terms for the Coast Guard; "leg," a paratroopers' term for nonparatroopers; and the non-Marine services' definitions of MARINE as "mostly Army rejects in Navy equipment" and MARINES as "many Americans running into never-ending shit." Good-natured (sometimes, not so good-natured) branch rivalry has long been a feature of military life; branch enlistic references this fact with venomous humor.

But probably the most common target for enlistic satire is the military itself, in all its bureaucratic wisdom and allegiance to protocol. In this regard the distinction between the O and E varieties comes through most clearly. Some officialese terms, in their ponderous precision, may seem to be unintentionally self-mocking. There is no such lack of intention when it comes to enlistic. In this bottom-up jargon, attacks on the institution, on policy, on individual missions, on incompetent officers, and even on the O jargon itself, are widespread and often hilarious.

An early and well-known example was the acronym SNAFU, which arose during World War II and quickly became a fixture of both military and civilian speech. It stands of course for "Situation normal—all fucked up" or, in a tamer version, "all fouled up." It was, and remains, the GI's ultimate comment on how the brass runs things, and its more recent appearance as FUBAR ("Fucked up beyond all recognition") only reinforces the common enlisted man's perception that "the wrong way" and "the Army way" are pretty much the same. That perception drives a host of other enlistic terms, such as "Puzzle Palace" for the National Security Agency, "Fort Fumble" for the Pentagon, "Rummy's Dummies" for Secretary of Defense Donald Rumsfeld's staffers during the first Gulf War; and the expression "echelons above reality" for the upper reaches of the military bureaucracy (Bay 2007a, 17). It is perhaps most cynically represented by the enlistic "hidden meaning" for USARMY: "Uncle Sam ass raped me today." This last one correlates well with Marines' identification of their branch's bureaucracy as the "Big Green Weenie" (Weems, chapter 7 in this volume).

Humorous disdain for the military is also evident in those enlistic terms—there are many of these—that play with O pomposity by creating satirical jokes that sound official. FOB, for example, is the official acronym for "forward operating base." In current enlistic, a "fobbit" is someone who never leaves the relative safety of such a base—a soldier who, like one of J.R.R. Tolkien's hobbits, prefers the comforts of a protected zone to the hazards of battle. In Vietnam, this same character was called a REMF, for "rear echelon mother fucker" (Clark 1990, 426). Alternate insults included "rear area pussy" and "house cat" (Reinberg 1991, 182, 180).

A similar tweaking of officialese can be seen in the fake acronym DILLIGAF. A civilian familiar with officialese might try to puzzle out the Pentagon meaning for that term, until he or she is told that it is enlistic for "Does it look like I give a fuck?" Another example is SIDPERS, which in the O vocabulary stands for "standard installation/division personnel system" but in the E vernacular means "silly idiots desperately pretending everything's running smoothly" or "SIDetrack personnel" (Bay 2007a, 43). So common were such in-jokes during Vietnam that enlisted personnel invented numbered lists of procedural signs, or "prosigns," each one keyed to a cynical catchphrase whose meaning could be evoked merely by uttering the number. "Number 7," for example, might stand for FUBAR, and "Number 21" for DILLIGAF (Cragg 1980).

Even when easily mocked acronyms are not in play, enlistic has its fun with the official jargon. In Vietnam, for example, villages were sometimes the subjects of "seek and destroy mission[s] in which American and South Vietnamese troops surrounded a village as South Vietnamese police searched for arms, guerillas, and political infrastructure [while] the villagers were provided entertainment and welfare and other services" (Reinberg 1991, 52). The O term for this was "Cordon and Search." The grunts doing the searching, alert to the "entertainment" aspects, called such an operation a "county fair." Similarly, when a patrol today is sent out with food and medicine to "win the hearts and minds" of a local population, those on patrol call it a "Gatorade run" (Bay 2007a, 20). If they go out in heavily fortified Hummers, the Pentagon says they are using "up-armored" vehicles (Glowka et al. 2008, 95). The boots on the ground, who are closer to reality, refer to such dubious protection as "hillbilly armor" (Bay 2007a, 24). Or consider, finally, two parallel terms for what civilians call a flashlight. The Department of Defense dictionary refers to it as (no kidding) a "flame field expedient," and further defines it as "a simple hand made device used to produce flame or illumination." The enlistic term for the "simple hand made device" is "moonbeam" (Bay 2007a, 31). In many other situations, there is an O and an E term for the same item or activity, with the latter one being figuratively "closer to the ground" than the former.

CODE-SWITCHING: THE O-E CONNECTION

The frequency with which O and E terms coexist brings to mind Ferguson's observation about "paired" vocabulary terms in diglossic communities, and such pairing is certainly one piece of evidence that military speech may be seen as a form of diglossia. The O-E distinction is like the H-L one, too, in that the split is standardized: Echoing Ferguson, we can safely say that the two milspeak varieties "exist side by side throughout the community, with each having a definite role to play" (1980 [1959], 232). As we've seen, the origin of the two varieties also follows Ferguson's H-L model: O is the product of command and management, while E emerges mostly from the ranks, among those who job is to follow, not deliver, orders. And finally, the O variety, like Ferguson's H, is a written jargon—the language of memos and reports—while enlistic, like Ferguson's L, is largely oral. E terms can be written down, of course, and collected in word lists. But they emerge from and flourish in oral folklore, not in the offices and war rooms of a paper-rich bureaucracy. So in several ways the case for military diglossia looks solid.

But there are complications. The most obvious one is that, whereas H and L varieties of a natural language typically inhabit strictly separated social realms, there is a great deal of overlap, and even of mixing, of O and E in the military environment. Ferguson makes it clear that, in true diglossic situations, the two varieties of the source language do not, or at least should not, mix. Each is appropriate to certain specified situations, and when they do mix—when L is used too formally or H too informally—there is a breach of linguistic etiquette that signifies cultural ignorance. This is not at all the case in the military realm. Here O and E are in constant contact with each other. There is no rule, formal or informal, that the two must never meet, and in fact, as several of the examples above indicate, enlistic is often a response to official pomposity. Some of its terms, like DILLIGAF or "echelons above reality," exist only because officialese exists. Military speakers, practicing what linguists refer to as code-switching, often employ O and E language in the same social situation—even in the same sentence.

We find a good example of this code-switching in a comment posted on Chazz Pratt's *Going Civilian* blog. "I need to top off the POV after I leave HQ, then change out of my Class A's and LPC's. It's about 10 klicks away so I'll stop by the BEQ before we link up and go get chow!" (July 11, 2011). Translated into everyday English, this means roughly, "I need to put gas in my car after I leave headquarters, then change out of my uniform. It's about ten kilometers away, so I'll stop by my apartment before we meet for dinner." It's not clear whether this is a naturally recorded utterance or whether the poster invented it for comic effect. In either case, though, it illustrates well the blending of O and E—along with some basic "civilian" talk—that is characteristic of military speech. In this example, "top off" is civilian slang, the acronyms are officialese, and "klick" and "chow" are enlistic. This is the way people in the armed services—grunts and officers alike—speak all the time. Even in hierarchical situations (like addressing a general), there is rarely an attempt to keep the language "pure" of enlistic, and even in very informal ones (like barracks bull sessions), soldiers may feel more comfortable with "HQ" than with "headquarters." So in this respect the Ferguson dictum must be modified: Not only do O and E exist "side by side," but their natural occurrence is a blend, a talking to each other.

A second way in which the O-E mix differs from the H-L standard is that neither of the military speech varieties is really a "native" language. In a diglossic Greek-speaking community, people would acquire the vernacular tongue, *dhimotiki*, as children, and only later learn the classical, or literary, variety known as *katharevusa* (Ferguson 1980 [1959], 234). This is the standard diglossic arrangement: L is the language of the streets, H of the academy. Not so in the O-E situation. Here both varieties must be learned, and even though one is associated with higher echelons and the other with boots on the ground, they are both learned in basically the same venue: boot camp or an officers' training facility. Both, moreover, are learned at the same time and from the same sources: a combination of jargon manuals, DI instruction, and peer conversation. The "native" language, for most US service people, is everyday English. Both O and E are artificial constructs that serve to build cultural competence in a novel situation. And they do that by operating together. The American soldier must learn both varieties, and he or she must be able to code-switch comfortably between them.

One final distinction between the Fergusonian and the military forms of diglossia. As we've seen, in traditional diglossia the High language is considered to be the more beautiful and prestigious language—the one in which complex ideas are related, sermons and lectures are delivered, and poetry is written. This is not a good description of the Pentagon's officialese. Its prestige, such as it is, comes from the fact that it is the language of officialdom and authority, not because of any supposed linguistic elegance. In fact, as many of the examples I've mentioned make clear, even though service people utilize officialese constantly, attitudes toward it are often blatantly disrespectful. Many enlistic terms directly ridicule officialese, either by coming up with homespun alternatives ("moonbeam") to its excesses ("flame field expedient") or by mocking its penchant for acronyms with fake abbreviations of their own.

Sometimes, moreover, enlisted personnel undercut the prestige of the bureaucracy's jargon by using so much of it that it mocks itself. One example of this reductio ad absurdum technique appeared in the 1987 movie *Good Morning, Vietnam*, a line from which is this chapter's epigraph. Here the disk jockey Adrian Cronauer, played hilariously by Robin Williams, employs seven acronyms in a single sentence: VP (vice president), VIP (very important person), PC (press conference), QT (quiet), VC (Vietcong), MIA (missing in action), and KP (kitchen patrol). He uses them to make a simple comment—"Let's keep the dignitary safe or we'll be in trouble"—sound both self-important and ridiculous. Some of the acronyms aren't military, but Cronauer's cleverness still shows that there is something laughable about the military's fascination with obfuscating letters. Like Pratt's blog post, the scene provides humorous evidence that the Pentagon's respect for formal jargon is not universally shared by its code-switching personnel.

CONCLUSION

In this chapter I've used Charles Ferguson's famous discussion of bilingualism to examine how the American military's "dual-language" system operates on the ground, and to ask to what extent the simultaneous use of officialese and enlistic can be considered an example of diglossia. After considering examples of both language varieties, and after seeing how they operate in tandem rather than in isolation, it seems clear that while the O-E language system shares important characteristics with Ferguson's H-L model, it also departs significantly from that model. Perhaps the most important difference is that the "High" and "Low" varieties of military speech are in constant communication with each other, and in fact function less as discrete tongues than as complementary registers of the same language system.

Certainly both are needed by military personnel, or at least understood to be so—and needed in equal measure, because each one serves a function that the other one cannot. Officialese clearly has a higher status among the bureaucracy—it's the official language, after all—and job success in the military requires detailed knowledge of its lexicon and tone. But social success in the military also requires a familiarity with enlistic. As a new recruit, one is drilled in officialese as the "proper" language; but one can still get drilled for incompetence in the "improper" tongue. A soldier who doesn't know that "Oh Dark Hundred" means "too damn early in the morning" is just as likely to be ridiculed by friends as an officer who refers to the BOQ ("bachelor officers' quarters") as his "apartment."

Generally speaking, officialese is the language of business and command, the variety in which the smooth (or not so smooth) running of the military machine is accomplished. Enlistic is the language of off-duty speech, the recklessly creative and often transgressive variety that allows its speakers access to an emotional release that official channels are unable to provide them. The machine might run without emotion, but people cannot. That is why, in assessing the relative importance of the O and E varieties, I would give them more or less equal status in terms of communicative competence and "personnel satisfaction."

Probably they could not survive without each other. In many military situations—not just dangerous ones—mission success depends both on orders being understood and carried out efficiently and on the people carrying them out being comfortable working together. Officialese is designed to facilitate the first requirement, and though one might question how effectively it does that, it's clear that having a standardized, universal jargon to describe almost everything might help to facilitate the following of commands. Enlistic, on the other hand, provides a social glue that enables personnel to bond with each other in tricky situations and, through the use of an insider speech, to share their feelings, often humorously, about what they are doing. Officialese enables quick and secure communication. Enlistic, as an oppositional complement, serves as a release valve for the tensions of the battlefield and frustration toward echelons.

In the military, as in any large organization, communication is a complex affair, requiring an attention to nuance and hierarchical positioning that cannot be managed as simply as in smaller social groups. The dual-register common language that we call military speech is an important regulating mechanism in that environment. Its two halves, officialese and enlistic, are equally important features of communicative competence. Each half, as in a natural diglossic community, has "a definite role to play." In the military, however, those roles seem a little more in touch, a little more interactive, a little more dialogic, than the roles of H and L in standard diglossia. Given that interaction, we may be justified in seeing the O-E system as a kind of "advanced" diglossia—a language accommodation in which institutional propriety is "softened" by slang and in which the needs of both action and emotion are creatively satisfied.

WORKS CITED

Bay, Austin. 2007a. *"Embrace the Suck": A Pocket Guide to Milspeak*. New York: New Pamphleteer.

Bay, Austin. 2007b. "Iraq's Battlefield Slang." *Los Angeles Times*, January 28.

Burke, Carol. 2003. "Military Speech." *New Directions in Folklore* 7.

Clark, Gregory R. 1990. *Words of the Vietnam War: The Slang, Jargon, Abbreviations, Acronyms, Nomenclature, Nicknames, Pseudonyms, Slogans, Specs, Euphemisms, Double-talk, Chants, and Names and Places of the Era of the United States Involvement in Vietnam*. Jefferson, NC: McFarland & Company.

Cragg, Dan. 1980. "A Brief Survey of Some Unofficial Prosigns Used by the United States Armed Forces." *Maledicta: The International Journal of Verbal Aggression* 4 (2): 167–73.

Department of Defense. 2011. *Dictionary of Military Terms*. http://www.dtic.mil/doctrine/dod_dictionary (accessed 25 February 2011).

Elkin, Frederick. 1946. "The Soldier's Language." *American Journal of Sociology* 51 (5): 414–22. http://dx.doi.org/10.1086/219852.

Elting, John R., Dan Cragg, and Ernest L. Deal. 1984. *A Dictionary of Soldier Talk*. New York: Charles Scribner's Sons.

Ferguson, Charles A. 1980 [1959]. "Diglossia." In *Language and Social Context*, ed. Pier Paolo Giglioli, 232–51. New York: Penguin Books.

Glowka, Wayne, Sarah Wyatt Swanson, Susan Presley, David K. Barnhart, and Grant Barrett. 2008. "Among the New Words." *American Speech* 83 (1): 85–98. http://dx.doi.org/10.1215/00031283-2008-004.

Kenagy, S. G. 1978. "Sexual Symbolism in the Language of the Air Force Pilot: A Psycho-analytic Approach to Folk Speech." *Western Folklore* 37 (2): 89–101. http://dx.doi.org/10.2307/1499316.

Orwell, George. 1984 [1946]. "Politics and the English Language." In *The Penguin Essays of George Orwell*, 354–66. New York: Penguin Books.

Reinberg, Linda. 1991. *In the Field: The Language of the Vietnam War*. New York: Facts on File.

Rich, George W., and David F. Jacobs. 1973. "Saltpeter: A Folkloric Adjustment to Acculturation Stress." *Western Folklore* 32 (3): 164–79. http://dx.doi.org/10.2307/1498382.

Roulier, Joseph B. 1948. "Service Lore." *New York Folklore Quarterly* 4 (1): 15–32.

Towell, Juile E., and Helen E. Sheppard. 1985. *Reverse Acronyms, Initialisms, and Abbreviations Dictionary*. Independence, KY: Gale Research Company.

6

Sea Service Slang
Informal Language of the Navy and Coast Guard

Angus Kress Gillespie

> *One of the qualities that ensures the sanity of a sailor is a sense of humor.*
>
> —W.A.B. Douglas, former navigation officer, Canadian Navy

One function of an occupational folk group's slang is to distinguish the members of the group from outsiders, and thus to nurture a linguistic cohesion among those "in the know." Carol Burke alludes to this fact when she comments that the military's "informal vocabulary" serves, at the most basic level, to distinguish members of the armed services from civilians (2004, 106). In the maritime services, this distinguishing element is perhaps most evident in the specialized lexicon used to name parts of a ship. Anyone who has seen popular films such as *The Caine Mutiny, Mr. Roberts,* or *The Hunt for Red October* will be aware that, in the US Navy, no ship has a front or back, floors, or a bathroom; it (or rather "she") has a "fore and aft," "decks," and a "head." That sailors utilize this vocabulary and landlubbers do not constitutes an important distinction between the two groups.

But the distinction between sailors and civilians is only one of many that are reflected in "sea service slang." Equally important differentiations are made between new recruits and older hands, seamen and petty officers, Naval Academy graduates and ROTC graduates, the Navy and the Coast Guard—these are among the many binary distinctions that are embraced by members of the seagoing services, sometimes as a means of humorously disparaging other members and sometimes as a way of grousing about the service itself. In this chapter, by focusing on some of the more colorful

DOI: 10.7330/9780874219043.c06   116

examples, I will show how the seagoing services' customary lingo—their informal rather than their official vocabulary—both advertises and reinforces the distinctions that are so much a part of their hierarchical culture.

My sources for this investigation include not only some of the standard published sources, such as Gershom Bradford's *Glossary of Sea Terms* (1946) and John Rogers's *Origin of Sea Terms* (1985), but also personal conversations with sailors and coastguardsmen as well as a variety of online jargon dictionaries. Online sources are of course notoriously variable in their reliability, but I would argue that, despite occasional inaccuracies, they remain useful for capturing additions to the slang lexicon that have not yet appeared in more scholarly venues. Where possible, I have also checked the reliability of questionable terms with my personal informants.

I offer the result of my investigation not as an exhaustive catalog but as a sampling focused on the theme of binary distinction. This theme, I admit, is only one of several that might shed light on the extraordinary richness of this occupational jargon. Burke, for example, correctly points out that military speech also provides its users an outlet for humor, for the relief of anxiety, and for the expression of frustration (2004); I'll be pointing to these elements too. In fact, my sample shows that, in an environment that can be alternately tedious and terrifying, sailors and coastguardsmen employ slang as a humorously derisive coping mechanism. What I am calling binary distinction functions as a major, but not exclusive, element in that language system.

FROM "RICKY" TO "GOB"

Let me begin my examination of Navy slang at what is the beginning for most enlisted men—Navy boot camp, also known as the Recruit Training Command at Naval Station Great Lakes, near north Chicago, Illinois. At boot camp an initial week of processing is followed by eight weeks of training, during which, instead of uniforms, recruits are issued "Smurf suits," a sort of blue jogging attire. Thus, the exposure to slang's discriminatory function begins with the recruit's obligation to dress like the small, blue fictional "heroes" of a comic-book series and an animated television show. Being required to wear a Smurf suit indicates that the new recruit is the lowest person on the food chain and a figure of ridicule. The older recruits, who get to discard the Smurf suits except for PT (physical training) view the new ones with pity, contempt, or both ("Surviving Military Boot Camp" 2011).

To further advertise their low status, recruits whose vision needs correction are prohibited from wearing contact lenses or their own civilian glasses. Instead, they receive official government-issued glasses with thick, hard-plastic frames and lenses. The pragmatic value of these glasses is that they are virtually indestructible. The symbolic value is that they identify the wearer as nerdy and sexually unattractive. Hence the slang terms for them: "BC glasses" ("birth control glasses") and "CGLs" ("can't get laids"). In a world where sexual prowess (or imagined prowess) carries high prestige, the slang designation announces the recruit's symbolic emasculation by more seasoned sailors ("Surviving Military Boot Camp" 2011).

Sexualized disparagement is also evident in the common nickname by which recruits are known: Ricky. This alliterative adaptation of "recruit" comes up in several different contexts. For example, "Ricky Rocket" is a boot camp energy drink made from soda, sports drinks, coffee, chocolate milk, and a large quantity of sugar, designed to keep the recruit awake despite grueling physical activity. "Ricky ninjas," according to a boot camp urban legend, are recruits who dress in black outfits and create nocturnal mischief. As further evidence of recruits' sexual low status, "Ricky boxing" is the common boot camp term for male recruits masturbating, while a "Ricky girlfriend" is the male recruit's right hand. With the increased presence of women in the armed forces, the relatively new term "Ricky fishing" has become a common euphemism for female recruits masturbating (Hoerger 2011).

Once they graduate from boot camp, young sailors are permitted to jettison the "Ricky" moniker and adopt the marginally less insulting one of "gob." The derivation of this early twentieth-century Americanism term is not clear, though some say it may refer to young sailors' tendency to gobble their food. (Rogers 1985, 80) In the old days a sailor might also be called by the inoffensive term "Jack Tar." The term may be explained by the fact that seamen would typically waterproof their clothing with tar before departing on a voyage (Rogers 1985, 178).

But the mildly disparaging "gob" is only one of many monikers that a new sailor may have to get used to. Beginners who are just learning the Navy way of doing things are subjected to barrage of personal insults. For example, a young sailor just out of boot camp might be called a "bubblegummer," a clear reference to his or her youth. Other terms are even more demeaning, such as the mock acronyms CUNT, meaning "currently unqualified naval trainee," and JAFO, meaning "just another fucking observer," referring to recruits who are so fresh in the fleet that they have not cleared any training. Still another is NUB, meaning "new useless body," referring to newly reported sailors—also called "newbies"—with no qualifications or experience. These sailors are usually given the most dirty and undesirable jobs, often called "shit work." For example, inexperienced sailors might be assigned to the mess deck or galley. But since they are not cooks, they serve the food, wash the dishes, and clean up. They may be called "mess bitches," a term that obviously implies derision and contempt and, less obviously, sexual vulnerability. Or they may be called "mess cranks," an arcane reference dating back to the age of wooden ships, when crews were called to mess by a hand-turned crank, or clacker, mounted on a mast (Tomazic 2011). Presumably new sailors were originally those who turned the crank; that they became known as "cranks" themselves is another good instance of comic disparagement.

One way in which new sailors are made aware of their lowly status is through the use of slang in hazing rituals. Capitalizing on their ignorance of jargon, old hands send them on fool's errands designed simultaneously to humiliate, initiate, and educate them. For example, the newbie may be told to locate and bring back a "boatswain's punch." He presents himself at this warrant officer's cabin, makes his request, and is immediately punched very hard by the veteran sailor. Or a veteran may task a new sailor to find a can of "bulkhead remover," which he searches for in vain until being informed that, bulkheads being ship's walls, they are never removed. Similarly, the "flight line" is the area on a carrier where the aircraft are made ready for flight. So a gullible new sailor might be told, "Go get me 100 feet of flight line from the crash shack"—the crash shack being a locker containing emergency response equipment (Hoerger 2011). Or he might be asked to retrieve a "gig line," which old-timers know is the invisible vertical line aligning a properly uniformed sailor's shirt seam with his belt buckle and trouser zipper ("Glossary of US Navy Slang" 2011).

In these cases, the arcane terminology describing mythical items serves at first a differentiating and then an incorporating function. This is also true of the jargon terms for actual items. The new sailor who refers to a battleship as a "boat" or to its deck as a "floor" is likely to be mocked by older hands for his ignorance of the proper terminology. The same would be true for the newbie who refers to the government-issued item on his head as a hat. The official Navy term for that item is "cover" (because it covers the head), while the common slang equivalent for the male head cover (figure 6.1) is "Dixie cup," from its resemblance to the disposable drinking cup.

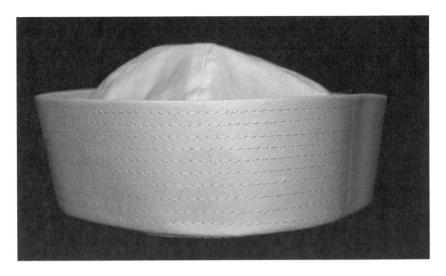

6.1. Dixie cup.

The uniform accompanying the Dixie cup is known as the "Cracker Jack," from its resemblance to the outfit worn by the child on the Cracker Jack box (Jones 2011).

"Dixie cups" are popular with sailors because they are inexpensive, foldable, and washable; the comic disparagement implicit in the term—and in the image of a sailor wearing a cup on his head—seems not to be a matter of particular concern. Female enlisted sailors wear a dressier headgear called a combination cover, or "combo cover," made up of a dark blue frame, a white slip cover, an elastic band that connects the frame and cover, and a patent-leather chinstrap ("Glossary of US Navy Slang" 2011). For both genders, the adoption of the unusual headgear—along with the corresponding slang terminology—differentiates the new sailor not only from civilians, but also from the Rickies recently left behind. At the same time, as he becomes familiar with the insider lexicon, his increasing fluency certifies him as worthy of inclusion.

MOVING UP THE LADDER

After surviving a period of hazing, a sailor may hope to advance in "rate." "Rate" is the enlisted man's equivalent of the officers' "rank." It indicates where the sailor stands in the chain of command and also his pay grade, or rate of pay. The Navy recognizes nine rate levels, with the first three (E-1 to E-3) assigned to apprentices, the second three (E-4 to E-6) to noncommissioned or "petty" officers (from the French word *petit*, for "small"), and the top three (E-7 to E-9) to chief petty officers. Sailors advance in rate through a centralized competition that uses a combination of factors, including having the required time in rate, a written examination, consideration of performance, and awards received. The idea is to make sure that, before being promoted, the sailor has shown leadership ability and the potential to assume increased responsibility (Bearden 1990, 187–189).

Ideally, to get ahead a sailor should avoid making mistakes or antagonizing superiors. In reality, this ideal is rarely achieved—which makes for an expansion of demeaning vocabulary. A sailor who is not too bright or who makes many mistakes, for example, may earn the unhappy sobriquet "a box of rocks." One who is not necessarily incompetent but who fails to follow orders in a timely and efficient manner may be identified as a troublemaker or as one resistant to authority. Given the power of the chain of command, sailors tend to follow orders dutifully. From time to time, however, one may object or talk back, and on occasion a junior sailor might even get into a shouting match with a chief petty officer. This unusual and ill-advised behavior is called "stepping out," as in stepping out of line or out of the prescribed arena of acceptable behavior. For this failure to observe the appropriate respect for hierarchy, a sailor may suffer nonjudicial punishment (NJP) ranging from a formal reprimand to reduction in rank, loss of pay, extra duty, or restrictions (Hoerger 2011).

Even worse than "stepping out" is "stepping in the shit." This term refers to a situation in which a sailor's mistake is so egregious that it comes to the attention of the commanding officer, who reprimands, or "chews out," the sailor on the spot. Similarly, "gaffing off" describes a situation in which a junior person ignores or purposely fails to show respect to a senior person. The term refers to the gaff, a long pole with a hook on the end used to land fish or (in the case of sharks) to keep them away. So just as one might use such a pole to push a shark away, a sailor might "gaff off" by procrastinating, neglecting an assignment, failing to salute, or failing to address an officer as "Sir" or "Ma'am." Infractions of this order may invite reprimands like, "When are you going to stop gaffing off and file those reports?" Those of an even more serious nature—for example, violations of the Uniform Code of Military Justice—may result in a bad conduct discharge (BCD), which sailors refer to wryly as the "big chicken dinner." The most serious violations might result in a dishonorable discharge (DD), or "duck dinner"

(Jones 2011). Getting either kind of discharge is a professional disaster, yet in translating expulsion from the Navy into a reward, the slang term may reflect sailors' conflicted attitude toward superiors, toward military justice, and toward the service itself.

By way of contrast, sailors who want to get ahead manage to either stay out of trouble or not get caught misbehaving. Even more important is a positive attitude toward work. The ambitious sailor is zealous about performing duties and always seems to be volunteering for more. In the old days of sail, such a sailor was called a "top sawyer," a crew member who was eager to please and to be useful. The term was borrowed from the old days of sawing logs or timber over a pit. Typically, there would be two men, a top sawyer standing above the timber and another man standing at the bottom of the pit. Traditionally, the sawyer holding the upper handle of the pit saw was the superior man—more skilled and more highly paid (Bradford 1946, 196).

The ambitious sailor must obey the rules and "learn the ropes," an expression carried forward from the age of sail not only in the sea services but in wider popular usage. The rigging in a large sailing ship could have ten miles of cordage divided into hundreds of separate parts, each with its own name and function. Thus "knowing the ropes" came to mean that seafarer's knowledge that distinguished the old hand from the beginner (Beavis and McCloskey 2007, 45).

At the same time, the ambitious sailor has to be careful not to distinguish himself so much that he makes his shipmates look bad by comparison: If he tries too hard, he may come across as a self-seeking flatterer. His shipmates might then say of him that he is "A.J. squared away," a term used to describe a sailor who always has a perfect shave, perfectly spotless and ironed uniform, and spit-shined shoes. The term "squared away" is used by all branches of the United States military services to describe one whose performance is above satisfactory. Such a sailor might also be called a "rate grabber," meaning that he has the goal of making rate quickly. This sailor might well antagonize his peers because they feel that he will try to advance at their expense. In the Navy, as in the civilian world, ambition is both admired and distrusted.

An ambitious seaman recruit (E-1) might quickly become a seaman apprentice (E-2) and then a seaman (E-3). But the big breakthrough is the jump to E-4, where he becomes a petty officer third class and is entitled to wear a distinctive insignia—a perched eagle with spread wings—on the left sleeve of the dress uniform above the rate chevron (figure 6.2).

6.2. The crow.

The affectionate slang term for the eagle is the "crow." (Mack, Seymour, and McComas 1998, 365). Given the poor reputation that crows enjoy—they are opportunistic scavengers, as opposed to soaring raptors—it's reasonable to read this slang term as a veiled attack on authority: It's the seamen's affectionate swipe at one of their number who, as an "A.J. squared away" or possibly a "rate grabber," has worked his way up the ladder to be their superior.

If the overachiever makes petty officer third class within his first term of enlistment, typically a four-year period, he may well decide to reenlist. If so, his shipmates will likely call him a "lifer," because he evidently loves the Navy enough to invest his whole life in it (Tomazic 2011); or a "cake eater," because most commands present such a sailor with a cake at the reupping ceremony (Hoerger 2011). Once he reenlists, he will be entitled to wear a service stripe, or "hash mark," to denote the length of his service. The Navy awards an embroidered stripe for each four years of duty, and sailors wear them diagonally on the bottom cuff of the left sleeve. It is not clear why the service stripes are informally known as "hash marks," but the term may have come from American gridiron football. There the hash marks are lines (about one yard long) used to mark every yard on the field between the five-yard lines (Bradford 1946, 85).

If all goes well, the eager beaver will pass smoothly through the ranks of petty officer second class and petty officer first class. Typically, it might take about fourteen years—sometimes more, sometimes less—for this person to make chief petty officer, and this represents the most significant promotion within naval enlisted ranks. It is based on the usual time in service, high evaluation scores, and specialty examinations. But there is the additional requirement of peer review, meaning that one becomes a chief only after review by a selection board. Chiefs are hierarchically—and one may also say psychologically—the Navy's equivalent of the Army's first sergeants, with the same reputation for toughness and expertise. Having attained the highest possible level of distinction for enlisted personnel, they are the target of both resentment and awe. This blended attitude explains the slang term for the chief's mess, a space on larger ships that is off limits to everyone but these crusty dignitaries. Sailors call it, with some affection, the "goat locker," that is, the domain of cranky old goats (Mack, Seymour, and McComas 1998, 366).

A similar ambivalence toward command is evident in sailors' terminology for officers, and especially for the ship's top officer, the CO (commanding officer") or captain. "Captain" is actually a Navy rank, between "commander" and "admiral," but by naval custom anyone who commands a ship is addressed as "captain," regardless of actual rank, while aboard in command. According to Naval Regulations, "The responsibility of the commanding officer for his or her command is absolute . . . The authority of the commanding officer is commensurate with his or her responsibility" (Mack, Seymour, and McComas 1998, 293). With such absolute authority, the captain sets the tone for leading and motivating his or her sailors. If the job is done well, the captain earns the respect and loyalty of subordinates. If the leadership is arbitrary or seemingly capricious, harmony and teamwork suffer, as they did notoriously, for example, aboard Captain William Bligh's HMS *Bounty*. Captains who are harsh disciplinarians are known as "sundowners" because they typically give their crews liberty, or permission to go ashore, only until sunset (Bradford 1946, 188). The resentment that sailors can feel at such restrictions is reflected in a joking quatrain of World War II vintage that refers to the captain's "gig," a small rowboat or motorboat that is reserved for his exclusive use when going ashore:

The lieutenant rides in a motorboat
The captain he rides in a gig

It don't go one goddamn bit faster
But it makes the old bastard feel big. (Tuleja 2011)

By contrast, a captain seen as more fair-minded is likely to inspire confidence and respect; such a leader is typically called the "old man." Clearly, this term evokes the image of a crew as a seagoing family, with the captain as the father—as in "Father knows best"—and the sailors playing the role of dutiful children. Other respectful terms for a CO include "skipper," from the Dutch *schipper*, for a ship's captain; and BMOS, for "big man on ship" (Hoerger 2011). In such terms, resentment at the absolute authority of a senior officer is modified by filial affection.

OFFICERS AND THE CHAIN OF COMMAND

Like enlisted personnel, officers have their own traditions, social norms, and slang terms for registering binary distinctions. Some of these evolve out of the different ways in which one may become a naval officer. Unlike enlisted personnel, all of whom begin their Navy careers at Great Lakes boot camp, those aspiring to officer status have a choice of starting points, most of which involve some sort of four-year college education. The most prestigious path to becoming a naval officer is to gain admission to the US Naval Academy in Annapolis, Maryland, the undergraduate college of the naval service, where the focus is on professional and leadership training in addition to academics. The Naval Academy has a mystique as the premier "leadership laboratory" of the Navy, a place that has graduated many "giants" over the years (Burke 2004, 126).

The Naval Academy experience is strict and stressful, especially for entering freshmen, who are called "plebes." This term was borrowed from ancient Rome, where Roman citizens were divided into patricians and plebeians, the latter being of the lower social order. Plebes must memorize all kinds of data on officer and enlisted ranks as well as all kinds of information on ships and aircraft. They must learn the NATO alphabet, from Alfa to Zulu, designed to avoid confusion in radio transmissions. They are required to recall information under pressure and—in a wonderful example of a constricted lexicon—they are allowed only five basic responses to questions from superiors: (1) "Yes, sir/ma'am"; (2) "No, sir/ma'am"; (3) "No excuse, sir/ma'am"; (4) "I'll find out, sir/ma'am"; and (5) "Aye, aye, sir/ma'am." Perhaps the most interesting of these responses is the fourth one, where plebes promise to learn the answer to a question that they are at the moment unable to answer. Ignorance, in other words, is no excuse. For someone planning to enter the esteemed ranks of the naval hierarchy, the answer to a superior's question is either instantly available or rapidly retrievable (*Reef Points* 2006–2007, 102).

Even at Annapolis, under the high pressure and rigor of the nation's most elite naval educational institution, slang expressions thrive. Some of these, like the shipboard expressions "bulkhead" and "head," are colorful jargon terms for common objects. For example, midshipmen—the academy term for "students," derived from a seventeenth-century term for junior officers—sleep not on beds but on "racks," the linens of which they are required to change once a week. If they are found sneaking a nap on the "bare zebra" (the black-and-white-striped mattress) while the bedding is being washed, they risk acquiring demerits and lowered standing (Harrison 2011). Other expressions reflect the midshipmen's low status, just as boot camp slang reflects that of Rickies. For example, no matter where a student comes from—and they come from every state and US territory as well as some foreign countries—his or her hometown is known as "Podunk" (*Reef Points* 2006–2007). The honorific title "anchor man" is given to the graduating student with the lowest class standing—still able to graduate with a bachelor of science degree and a commission in either the Navy or the Marine Corps (*Reef Points* 2006–2007, 130). And the academy itself—the "leadership laboratory" for giants—is known with affectionate mockery as "Canoe U" (Burke 2004, 108).

The Navy's need for officers cannot be met by the Naval Academy alone. Hence there are other paths to a naval commission. One way is to take up the Reserve Officers' Training Corps (ROTC) program while attending a regular college or university. Aspiring officers who choose this option normally take naval science courses along with their regular curriculum, and they attend at least one summer training session. A third path is to participate in the Officer Candidate School (OCS) program. This is a twelve-week program for college graduates who train at Naval Station Newport in Rhode Island. ROTC and OCS graduates have performed honorably in numerous naval engagements, yet Naval Academy graduates still have trouble taking such easily minted officers seriously. They refer to them as "Shake 'n' Bakes," Nestle's Quiks," or "Redi-Whips" (Burke 2004, 108–9); these terms are modern spins on the older pejorative term "90 day wonders," which has been used to refer to OCS officers since World War II.

Once they are commissioned and serving as officers, there is no official distinction between "Canoe U" graduates and other officers. Academy graduates are known as "ring knockers" because they display their graduation rings proudly (Mack, Seymour, and McComas, 1998, 369) and may in officers' meetings tap them quietly on the table to elicit support from fellow graduates. They may or may not have a leg up on the other officers. But all officers follow the same career path from ensign (the lowest commissioned rank) through the junior ranks (lieutenant junior grade through lieutenant commander) to the senior ranks (commander to the various admiral levels). Few officers advance beyond captain (the highest rank below admiral), and all begin as ensigns.

These young officers, equivalent to Army second lieutenants, wear one gold stripe on the sleeves of their dress uniforms. The decoration marks them as simultaneously senior to all enlisted personnel and the lowest-status officers on board—a strange combination of importance and impotence. As the most junior officer on board a surface ship, an ensign will be identified—by officers and seamen alike—as also the most ignorant. As a result, whether he likes it or not, other officers may call him "George," also spelled JORG, meaning "junior officer requiring guidance," or JORGE, meaning "junior officer requiring general education." A former chief petty officer told me that such officers would sometimes be obliged to wear oversized ensign bars (insignia) engraved with the jocular title "George" (Jones 2011). The term reflects the fact that new ensigns, like the eponymous hero of the expression "Let George do it," are likely to be given tasks that no one else wants: As the lowest officers in the hierarchy, that is their lot. The most senior of these "lowest," by the way, is sometimes, with sardonic gravity, called the "bull ensign" (Tomazic 2011).

Ensigns have much to learn linguistically as well. Just as Naval Academy plebes are subject to a constricted range of responses to superiors' questions, new ensigns must also observe a restricted vocabulary when conversing with fellow officers. Service etiquette has many carefully prescribed, almost jesuitically subtle, formulas for officers to communicate with each other. For example, seniors "call" or "direct" attention to something; juniors must "invite" seniors' attention. Seniors "suggest" that something be done; juniors may only "recommend." Seniors "direct" juniors to act; juniors "request." Juniors, both sailors and Marines, acknowledge an order by saying, "Aye, aye." Seniors acknowledge information conveyed by juniors by responding, "Very well." Under no circumstance should a junior officer say "Very well," even with "sir" tacked on to the end (*Other Traditions* 2001). The formulas are meant to make clear in every exchange who is up and who is down in the chain of command.

Typically, after two years of service, an ensign is eligible for promotion to the next rank, lieutenant (junior grade), pronounced "jay-gee." These officers wear a stripe and a half on their dress uniforms, but they are still very much junior officers. As such, on a typical surface ship they are assigned to junior officers' quarters, or "steerage," so called because in the old days the tiller projected into their compartment, located in the aft of the ship (Rogers 1985). Junior naval officers are subject to being ordered to carry out a variety of undesirable tasks. Candidates for this dubious honor are typically selected by the executive officer (second in command to the CO), according to a principle humorously called LOST, or "line of sight tasking," meaning that the senior officer gives the job to the unlucky junior officer who first walks across his path (Hoerger 2011).

Assuming a smooth career path, an officer might next pass through the ranks of lieutenant, then lieutenant commander, and then, after fifteen to seventeen total years of service, be eligible for promotion to commander. Commanders are the first level of the Navy's "senior officers," who have greatly augmented privileges. Notably, Navy commanders wear "scrambled eggs," the slang term for the gold, leaf-shaped braiding worn on the visors of their caps (figure 6.3).

The term is in common usage, especially with regard to admirals, who wear two rows of such braiding (Tomazic 2011). Wearing scrambled eggs is a true honor—a mark of professional achievement—yet the term may also be seen as humorous (and perhaps unconscious) disparagement. In a social arena where spit-and-polish dress signals excellence, a principal index of senior officer status is a decoration referred to as food on the uniform. The implicit critique of authority becomes all the more pointed in light of the fact that, to Navy men from the 1940s onward, powdered eggs were considered, in the words of one World War II officer, "not fit for humans to eat" (Anderson 2002). In this term, again, we can see how folk usage puts a comic, slightly mocking spin on an official distinction.

ALL THE SHIPS AT SEA

Informal usage also governs the terminology with which naval person-nel describe their ships—the vessels that are in fact their only protection

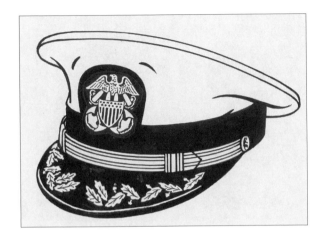

6.3. Scrambled eggs.

from the sea, but that are often described in terms that mirror the humorous disparagement with which they describe each other.

When most of us think of naval vessels, we usually think of large surface ships with many guns, including battleships and cruisers. In the old Navy, these large warships were sometimes called "ships of the line," meaning that they were capable of taking a position in the first line of defense or offense (Bradford 1946, 167). Battleships and cruisers are no longer the most important surface ships. Instead, ever since World War II, the aircraft carrier has taken on that role. Aircraft carries are warships designed to deploy and recover aircraft, function as seagoing air bases, and project national power worldwide. World War II slang for them—heard relatively rarely today—was "flattops," a reference to their flat flight decks (Tomazic 2011). The nickname also echoes the term "flat top" for the short-clipped haircut style that has long been popular with military personnel.

Since aircraft carriers are so important, they are normally placed at the center of a battle group, with a ring of outer, smaller ships for protection against enemy intrusion. Key among the outer ring of ships are the destroyers. They are fast and maneuverable, yet thinly armored, which gives them the comically disparaging nickname "tin cans" ("Glossary of US Navy Slang" 2011). Also useful in protecting the aircraft carrier are the frigates, which are sometimes called "missile sponges" because, with limited air defense, they are stationed on the outer ring of the battle group, where they are most likely to be hit by enemy fire; sailors sometimes describe their mission as "torpedo catching" (Jones 2011). Destroyers and frigates may also go by the slang term "small boys," to distinguish them from the "big-boy" battleships

6.4. Collision at sea.

and carriers (Mack, Seymour, and McComas 1998, 38). Thus the slang used to describe fighting ships is a gently mocking alternative to the official vocabulary. While a "destroyer" suggests power and invincibility, a "tin can" may be at the mercy of the enemy and of the waves. Slang here reflects a sailor's knowledge about the perils of Navy life that the official terminology obscures; it may also serve, through comic belittling, to relieve anxiety (Burke 2004).

Another example of this process may be observed in the slang terms applied to the badge worn by an officer of the deck, or OOD. This officer is stationed on the bridge, or command center, and in the absence of the captain or other senior office is responsible for the vessel's navigation and safety. To gain this position of authority and responsibility, an officer must undergo rigorous training in shipboard engineering, damage control, and quality maintenance. Upon completion of this training, he or she is awarded a gold badge that depicts waves breaking before the bow of a ship (figure 6.4).

This badge, worn proudly above the officer's ribbons, is known officially as the surface warfare officer insignia (Mack, Seymour, and McComas 1998, 38). In common parlance, though—one officer told me off the record—the badge is referred to as "collision at sea," because it looks as if the ship is about to strike the observer. Another officer jokingly told me that the insignia is called "water wings," from the similarity of the breaking waves to naval aviators' golden wings insignia. Both terms, like the term "crow" for the petty officer's insignia, provide a humorous sardonic response to the official service story.

Submariners have their own slang expressions that set them apart in interesting ways from the sailors on surface ships, which submariners call "skimmers" (Jones 2011). For example, while the surface Navy calls all of its vessels "ships," submariners call theirs "boats." Unlike the Navy in general, which has an elaborate classification system for its ships, submariners are fond of saying that there are only two types of vessels: "submarines" and "targets." Other expressions reflect the fact that the role of the "silent service" is generally not to engage with but to avoid the enemy. A crew member of a ballistic missile submarine is described, humorously but respectfully, as a "chicken of the sea," meaning that the mission is to avoid detection by whatever means necessary. Submariners often say, wryly, "We hide with pride" (Donne 2011). Because the undersea platform is designed for massive retaliation against an aggressor—an outcome that no one wants and everyone tries to avoid—the downside of life on board a ballistic missile submarine is that "nothing ever happens." As one submariner once told me, "Our job is drilling holes in the ocean." Like many other slang expressions, the remark reveals both frustration and pride.

COAST GUARD SLANG

Let me now turn to the slang of the United States Coast Guard (USCG)—described as a maritime, military, multimission organization. It was formed in 1790 by Alexander Hamilton, secretary of the treasury, to enforce his program of protective tariffs and to put the new nation on a sound financial footing. Unlike the other four armed services, it falls under the Department of Homeland Security rather than the Department of Defense. The official motto of the Coast Guard is Semper Paratus, or "Always Prepared."

Many of the Navy terms I have discussed also circulate among coast-guardsmen, but there are other slang expressions that have a distinctive Coast Guard flavor. For example, there is the classic slogan: "You have to go out, but you don't have to come back." This was the informal motto, never officially sanctioned, of the United States Lifesaving Service, one of the precursors of the Coast Guard. Today, it's a familiar saying among USCG helicopter rescue teams. Even though today, under extremely adverse conditions, a rescue team can be ordered to stand down, the willingness to risk one's life for others remains a core value of the Guard, and it is well reflected in this stock expression (*Reef Points*, 2006–2007, 128).

There is a long-standing rivalry between the Navy and the Coast Guard, some of it generated by the tremendous discrepancy in the size of the two branches. As of this writing, the Coast Guard has about 42,000 men and women on active duty, compared with the US Navy's more than 381,000. Because the Coast Guard has many small boats and only a few rather small ships, called cutters, Navy personnel often mock the Coast Guard as the "knee-deep Navy" or "Uncle Sam's Confused Group" (USCG). Navy personnel may refer to coastguardsmen as "shallow-water sailors," "puddle pirates," or "mud ducks." They may take delight in reciting the old joke: "Why do you have to be six feet tall to join the Coast Guard?" Answer: "Because if your cutter sinks, you can wade ashore" (Krotz 2011).

Coastguardsmen are as alert to the invidious distinction as their Navy counterparts, and in response they sometimes explain that they are too busy simply doing their jobs to take offense at these Navy slurs. They say also that the Navy basically has no peacetime mission except to train over and over again for a war that everyone hopes will never happen. The Coast Guard, on the other hand, is busy every day saving lives, boarding fishing vessels, inspecting merchant vessels, maintaining aids to navigation, interdicting migrants, and so forth. But, when provoked, coastguardsmen do have a pointed comeback to Navy jabs: "If it weren't for us, you would have ugly kids." That is, coastguardsmen are romancing unfaithful Navy wives while their husbands are away on extended deployments (Kennedy 2011). The joke is an interesting variant on the "Jody" tradition (see chapter 4 by Burns in this volume), in which the cuckolding of the absent serviceman is accomplished by another serviceman.

Coastguardsmen also display rivalry among themselves. Although all Coast Guard personnel are on the same team, there are inevitable intraservice rivalries stemming from the fact that Coast Guard vessels are loosely grouped into three distinct fleets, differentiated by the color schemes of their hulls. Although any Coast Guard vessel may be assigned to any task, in general black hulls are for working cutters, those that maintain aids to navigation; white hulls are for maritime law enforcement; and red hulls indicate icebreaking vessels. Rivals sometimes call the red-hulled cutters "polar rollers," reflecting the way in which these vessels "roll up" onto ice in order to break it (Uronis 2011). White-hulled cutters, which function like seagoing cops on a beat, are sometimes mocked for "riding around and wasting fuel," while their shipshape appearance elicits the jibe that their unofficial motto is "Don't scratch my white paint." Black-hulled cutters take on the heavy and messy work of repairing and replacing buoys, which gives them the tagline "the workingman's fleet." Such mildly disparaging remarks tend to reflect not antagonism but good-natured rivalry: As one white-hulled cutterman said to me, "We tease them, but we admit they earn their pay" (Leahy 2011).

I've described the way in which slang usage provides humorous "correction" for the names of Navy ships. The same satirical undercutting goes on in the Coast Guard as a response to the service's alphabetical code system. The codes for all Coast Guard cutters begin with the prefix "W" to distinguish them from their Navy counterparts: The remainder of the code indicates their operational specialty. For example, WMEC stands for "medium endurance cutter" and WHEC for "high endurance cutter." In response to this official terminology, those on board high endurance cutters are fond of the quip "We have to eat chicken," while those on medium endurance cutters say, "We must eat chicken." In both cases, the crew is making fun of the service's policy of serving poultry in different disguises to hold down food costs.

Joking jargon has also evolved in response to the service's code designations for icebreakers: WAGB and WTGB. There is no official explanation for these acronyms, although a reasonable guess might be that they stand for "arctic glacial breaker" and "tug glacial breaker." To their crews, however, the big polar breakers are known as "wandering arctic garbage barges," appropriate because of the huge amount of trash that accumulates during a voyage and because the scientists on board dictate the apparently aimless course. Likewise, crews on the smaller icebreaking tugs might explain their absence on a given day by saying, "Went to get beer" (Kennedy 2011).

The official names of individual Coast Guard cutters also lend themselves to language play. The cutter *Active*, for example, carried the nickname "Inactive" because in the later 1970s and early 1980s she spent most of her time moored due to problems with the turbines. The *Campbell* was commonly known as "Soup Can," the *Courageous* as "Outrageous," the *Evergreen* as "Evergone" due to her extended deployments away from home, and the small icebreaking tug *Penobscot Bay* as "Peanut Butter Bay." In all of these cases—and they are only representative—informal usage chides the dignity of official terminology and maintains a line between formal protocol and how sailors actually speak ("Cutter Nicknames" 2011).

The rivalry among the black-, white-, and red-hulled communities is tame compared to that between the seagoing personnel and the people ashore, who may be referred to as "coffee-swilling, donut-eating, sand peeps" ("peep" is a slang term for the sandpiper, which spends its life on shore). There is also a profound rivalry between the seagoing personnel and those in aviation. The larger Coast Guard cutters have helicopter crews on board. These crews include pilots as well as maintenance and avionics technicians. The aviation crews are regarded with a mixture of envy and hostility. In the view of seagoing personnel, the aviation community has unusually easy duty. The ship's crew members have the burden of watch standing. They have to operate the ship continuously around the clock, so they tend to have extended hours in addition to their regular hours, resulting in interrupted sleep patterns. One can imagine their annoyance when they see the aviators taking daytime naps and watching movies in the evening. So the aviators are derisively called "rotor heads," "wing nuts," or "cloud jockeys" (Kennedy 2011). Such terms provide another example of how informal usage can shed light on the complexity of sea service life in ways that official pronouncements and protocol may not.

CONCLUSION

Slang has been referred to as a "glaring misuse of register," a form of speech that lowers "the dignity of formal or serious speech or writing (Dumas and Lighter 1978, 14–15). Those who champion the "officialese" of the armed forces as an appropriate formal register and who denounce the more rowdy "enlistic" (see chapter 5 by Levy in this volume) would no doubt agree. Slang does undercut the dignity of officially sanctioned speech, and it lends itself to a directness of presentation that, in a hierarchical context like that of the Navy, can be not only troublingly coarse, but humorously disruptive of decorum.

But disruption, of course, is one of the pleasures of slang. To be able to chide a preening social climber as a "rate grabber," to smile knowingly about one's vessel as a dry-docked "Inactive," to see a chief's badge as a "crow" rather than an eagle—these are minor swipes at propriety that, in the military context, may afford the powerless a vestige of temporary control. Even if that control is illusory, it may be psychologically comforting to common seamen and junior officers, and may thus "relieve" not just anxiety but potential refractiveness. To quote the Canadian naval officer whose comment serves as this chapter's epigraph, such humorous ribbing of the powers that be may even help to ensure a sailor's sanity (cited in Jenson 2000).

Slang may thus serve almost a conservative, reassuring function. Without the occasional release that is afforded by such humorously disruptive displays, sailors and coastguardsmen—often standing a double watch against boredom and peril—might well feel more frustrated, more anxious, about their condition than they do. So even crude examples of slang may function, psychologically, to keep the ship running smoothly.

Besides, the "corrective" function is only part of the story. Carol Burke and others are surely right that such informal speech expresses anxiety and frustration. But in the sea services—as no doubt in the other services—those sentiments mingle with dedication and pride. The examples I've discussed reveal this as well. Expressions like "goat locker" and "old man" have an honorific ring to them, even when the authority so referenced is acknowledged grudgingly. Derogatory monikers like NUB and JORGE, even as they ridicule newcomers to the ranks, imply that membership in those ranks is worthy of admiration—and that there is pleasure associated with the duty of educating the ignorant. When a submariner boasts that he "hides with pride," he is doing more than complaining about lack of activity; he is registering a sense of accomplishment with a job well done. That this is done with self-deprecating humor in no way diminishes its effectiveness.

Anyone who has heard members of the armed forces speak of their service will acknowledge that their reflections commonly display a mixture of irritation and pride. That mixture is certainly evident in sea service slang, and its presence may alert us to the irony that people in uniform are at once the most vocal critics of military procedure and its staunchest defenders. To those who either despise or lionize the military, this attitudinal complexity may not be evident. I hope the examples of maritime slang discussed here may remind us of how central it is to the service experience.

ACKNOWLEDGMENTS

I am grateful to Eric Eliason and Tad Tuleja for encouraging me to write this chapter. Several people helped me along the way. On the Navy side, Captain David "Fuzz" Harrison, USN, and Commander Edward Hogan, USN, offered me valuable advice. On the Coast Guard side, Captain Linda L. Fagan, USCG, was immensely helpful, as was Charles Rowe, public affairs officer for USCG Sector New York. Especially generous with their time were BMC Troy M. Katz, USCG, and Lieutenant Robert Lee Kennedy, USCG, both of whom had wonderful stories. Finally, I would like to thank my student Victoria Verhowsky, who helped compile, and then sort, huge lists of words from which I chose the selections discussed in this chapter.

WORKS CITED

Anderson, P. Lanier, Jr. 2002. In interview with Patricia B. Mitchell, "WWII Navy Food Remembered." http://www.foodhistory.com/foodnotes/leftovers/ww2/usn/pla (retrieved 7 July 2011).

Bearden, Bill. 1990. *The Bluejackets' Manual.* 21st ed. Annapolis, MD: United States Naval Institute.

Beavis, Bill, and Richard McCloskey. 2007. *Salty Dog Talk: The Nautical Origins of Everyday Expressions.* Dobbs Ferry, NY: Sheridan House.

Bradford, Gershom. 1946. *A Glossary of Sea Terms.* New York: Dodd, Mead, & Company.

Burke, Carol. 2004. *Camp All-American, Hanoi Jane, and the High-and-Tight: Gender, Folklore, and Changing Military Culture.* Boston: Beacon Press.

"Cutter Nicknames." USCG History Program. http://www.uscg.mil/history/webcutters/ Cutternicknames.pdf (retrieved 29 May 2011).

Donne, John. 2011. Lt. Commander, USN. Personal interview, July 21.

Dumas, Bethany K., and Jonathan Lighter. 1978. "Is Slang a Word for Linguists?" *American Speech* 53 (5): 14–5.

"Glossary of US Navy Slang." 2011. *Wiktionary.* http://en.wiktionary.org/wiki/ Appendix:Glossary_of_U.S._Navy_slang (retrieved 15 May 2011).

Harrison, David. 2011. Captain, USN. Personal interview, July 21.

Hoerger, Jeffrey P. 2011. Lt. Commander, USNR (Retired). Personal interview, July 26.

Jenson, Latham B. 2000. *Tin Hats, Oilskins, and Sea Boots: A Naval Journey, 1938–1945.* Toronto: Robin Brass Studio.

Jones, Charles. 2011. Chief Petty Officer, USN (Retired). Personal interview, July 26.

Kennedy, Robert Lee. 2011. Lieutenant, USCG. Personal interview, May 6.

Krotz, Troy M. 2011. Chief Boatswain's Mate, USCG. Personal interview, May 6.

Leahy, Scott. 2011. Chief Boatswain's Mate, USCG. Personal interview, July 21.

Mack, William P., Harry A. Seymour Jr., and Lesa A. McComas. 1998. *The Naval Officer's Guide.* Annapolis, MD: United States Naval Institute.

Other Traditions of the United States Naval Services. 2001. http://mysite.verizon.net/vzeo-hzt4/Seaflags/customs/trads.html#phrase (retrieved 15 May 2011).

Reef Points: The Annual Handbook of the Brigade of Midshipmen. 2006–2007. Annapolis, MD: Naval Institute Press.

Rogers, John G. 1985. *Origins of Sea Terms.* Mystic, CT: Mystic Seaport Museum.

"Surviving Military Boot Camp." 2011. *About.com* (2007–). http://usmilitary.about.com/ od/navyjoin/l/aanavybasic1.htm (retrieved 15 May 2011).

Tomazic, Rocco. 2011. Commander, USN (Retired). Personal interview, July 26.

Tuleja, Tad. 2011. Private correspondence, March 4.

Uronis, Richard. 2011. Master Chief, USCG. Personal interview, July 21.

Part III
Belonging

7

Taser to the 'Nads
Brutal Embrace of Queerness in Military Practice

Mickey Weems

Queers and the military have made strange bedfellows in American history. The relationship resembles that of two lovers in a forbidden, secret affair in which one partner is abusive, and neither of them can say good-bye. The following is an account of the uneasy dynamics of homophobia in military men's individual and team identities. It includes analysis of *fucker* and *pansy* antiheroes, historical factors leading to the imposition of official silence by military bureaucracy, unofficial folk speech with transgressive homoerotic humor, and outrageous folk performance-as-resistance videos in the face of perceived institutional oppression, including humorous resistance to Don't Ask, Don't Tell by Straight men as they stood up for their Gay brothers. It concludes by examining the videos that, since the second invasion of Iraq, have lampooned both Queers and official military paranoia concerning all things Queer, and thus tacitly show support for Gay personnel while simultaneously making fun of them.

IT'S COMPLICATED

Until 2011, homosexuality was officially deemed incompatible with military service. Nevertheless, the US military quietly bunked with closeted homosexuals right from the start. Sometimes the illicit relationship was unspoken but accommodating, especially when fighting men were in short supply. The first national drillmaster, Baron von Steuben, for example, was renowned for his military bearing, intelligence, discipline, humor, love for his

DOI: 10.7330/9780874219043.c07   139

troops, and love of men. He was never prosecuted for sodomy due to his crucial role in winning the War of Independence. At other times, though, homosexual personnel were ferreted out, humiliated, and dishonorably discharged.

Tensions were aggravated in the 1970s when servicemen accused of
doing wrong with their brothers-in-arms publicly rejected stigma attached to
homoerotic-romantic love. Acceptance of LGBTQ (Lesbian, Gay, Bisexual,
Transperson, Queer) people in the late twentieth century inspired the military to play a convoluted game of erotic hide-and-don't-seek by implement-
ing the linguistic fiction known as Don't Ask, Don't Tell (DADT) in 1993.
DADT did not officially forbid or condemn homosexuality just as long as
homosexual behavior and Gay identity remained unspoken and unseen. But
the military bureaucracy could not resist playing the voyeur. Gay service
personnel were "asked" and discharged when they "told," even if the telling
was inadvertent, such as an overheard conversation, a chance observance of
an off-duty and off-base encounter, or covert surveillance of mail and e-mail
by their superiors.

DADT was little comfort to Gay service personnel because it was not
created for their benefit. Rather, DADT's true purpose was to allow the
military to distance itself from its ongoing affair with queerness, a sordid
fetish that not only punished homosexual men but brought scandal after
scandal against the military because of significant blind spots in the rhetoric of homophobic masculinity. These scandals include a YMCA sex sting
gone wrong (1919), a recruiting video featuring the Village People (1979),
congressional debate on a Gay men's dance party as antimilitary (1996),
publication of the Fag Bomb photo just after 9/11 (2001), and revelation of
a secret Gay Bomb program (2004).

Officially, the military wanted nothing to do with homosexuality before
the end of DADT. But unofficially, homosexuality has been a hot topic for
military men since at least the mid-1800s as they insulted each other and
bonded together through transgressive humor. The term *bugger* (homosexual) was used by soldiers in the Civil War, often as a derogatory term for an
officer by enlisted men (Wright 2001, 45). Homophobic-erotic terms such
as *grab-ass* (referring to homosexual play, and signifying disorderly conduct
and silliness) can be traced at least to World War II (Rottman 2007).

Hollywood brought such language out of the barracks and into movie
theaters across the nation. Since the 1980s, homophobic-erotic language
and behavior have been hailed by script writers and directors as a hallmark
of the real, uncensored military in several war films: a male Army soldier

engages in quasi-fellatio by sucking hashish smoke from a rifle barrel blown into his mouth by another soldier (*Platoon*, 1986), a Marine drill instructor makes constant references to homosexual acts when berating recruits in boot camp (*Full Metal Jacket*, 1987), male Marines mimic sex with each other to scandalize embedded reporters (*Jarhead*, 2005), and a female SEAL trainee, rising to the challenge of the all-male domain, tells a misogynistic captor in POW simulation, "Master Chief, suck my dick" (*GI Jane*, 1997). All of these scenarios became iconic rather than problematic because audiences loved them for being genuine, while the military-as-institution made it clear it had nothing to do with their production.

Infatuated young men learn military homophobic-erotic language from movie dialogue before they go into training, then create new language once they are in the armed forces, generating movie-certified and real-life folk speech that script writers love to incorporate in their films. In the twenty-first century, military personnel have been making videos of their own, some with humorous homoerotic references performed through parody and dance. Some punitive and negatively framed actions concerning homosexual men, however, have caused great embarrassment for the US military, the most instructive example being the Fag Bomb.

HIJACK THIS

On October 11, 2001, one month to the day after al-Qaeda launched attacks on American soil, a bomb was prepared on an aircraft carrier for deployment in Afghanistan. On it was a message: HIGH JACK THIS FAGS. A photograph of the weapon made its way to the Associated Press to inspire and amuse the folks back home. But the Gay and Lesbian Alliance against Defamation (GLAAD) was neither inspired nor amused. AP pulled it from the news and apologized, as did the Department of the Navy.

Members of al-Qaeda were called "fags" because their organization penetrated American defenses and killed Americans. Success in humiliating the United States demanded retaliation with bombs, bullets, and battalions to penetrate the new enemies; calling them *fags* was verbal flourish. But *fags* is a cuss word, not fit for official military discourse, and, outside of legal language prosecuting homosexuals, there was zero tolerance for any language concerning homosexuality, pro or con. The Fag Bomb photo was an embarrassment to the Navy not because it insulted the Gay community, but because it referenced Gays and, on top of that, used a cuss word.

Homophobic-erotic men's folk performance is common in military men's private discourse. But it has proven disastrous in public discourse representing the military-as-institution. What is in between man and institution, the *team*, becomes contested ground when homophobic-erotic themes are voiced in private, exposed to public scrutiny, and then silenced, which is exactly what happened with the Fag Bomb. Produced by the crew of the aircraft carrier, it was a team effort that would never have caused a stir if the picture had been limited to the crew. The Fag Bomb photo shot out of the private world of military personnel, landed in the headlines, and blew up in the Navy's face.

ANATOMY OF CUSSING

Men's military folk speech has plenty of negative references to male homosexuality, reflecting its fetishization in the general population by Straight American men when they insult each other man to man, and insult each other's sports teams fan to fan.

Insults occur within the frame of one man *cussing* (using obscene language against) another man. Cussing somebody is an immediate body-to-body activity that is more visceral than cerebral. Its purpose is to create a ritualized frame for battle through verbal transformation that resembles incantation: If I call you a fag and you do not respond with sufficient resistance, you become what I invoke. This is the underlying rationale of the Fag Bomb, what the folk call *fighting words*, often accompanied with appropriate gestures, stances, and facial expressions.

Analysis of cussing reveals a theme in sexual discourse for men: penetrating is good, being penetrated is demeaning. A penetrated man is a defeated man, an effeminate man, a weakling, a laughingstock, and this notion goes back thousands of years. Biblical stories of attempted male rape in the corrupt cities of Sodom (Genesis 18:16–19:29) and Jebus (book of Judges 19:1–30) illustrate that one way of dealing with the male stranger is to shame him publicly, sexually, and against his will. The association between man-on-man rape and effeminacy is grounded in the unspoken notion of men's contested anal virginity. A man penetrated by another man is no longer a real man, much as a woman whose vagina is penetrated by a man's penis is no longer a virgin.

A tendency to humble the outsider by means of homosexual rape is with us in the twenty-first century. Folk speech in sports (such as "FUCK

MICHIGAN" and "MICHIGAN SUCKS" T-shirts sold near Ohio State University in observance of the annual OSU-Michigan game) glorifies verbal description of forced sexual penetration, that is, festive rape. "Fuck Michigan" and "Michigan Sucks" are battle cries, examples of the close proximity of football to the military. Although cussing the opposing team is theoretically comedic, it occasionally results in violence because it includes fighting words that are the verbal kin of such everyday taunts as "Fuck you," "You suck," "Suck my dick," and the Gay male sadomasochistic phrase that appears to have been lifted directly from Leather sexuality: "I'll have your ass in a sling." (A sling is a leather seat suspended from chains, designed to position a man seated in it so that his anus may be more easily penetrated in face-to-face sexual intercourse. The Leather community is dedicated to the practice of bondage/discipline/dominance/submission/sadomasochism, or BDSM.)

These curses constitute verbal sexual assault and an invitation to fight, usually occurring in recreational contexts such as sports arenas, festivals, and bars. Most of the time, it is not play-fighting, that is, a mock battle with no intention to harm. This form of cussing/cursing has the potential for real violence. Fighting words are complemented by phallus-mimicking gestures that signify unwanted penetration: "shooting a bird" (aggressively displaying the middle finger turned upward by itself) and the aptly named "up yours" (a fisted hand thrust upward while the opposite hand comes down upon the flexed bicep). The meaning of both is "I thrust my erect penis up your ass," equivalent to the laconic "Fuck you."

The gist of these phrases and gestures is this: My erect penis is a weapon of destruction and humiliation. Its purpose is neither pleasure nor procreation. Whatever joy I may get from fucking my enemy is derived from my enemy's distress. As such, I am not a fag, but I can turn the insulted into a fag against his will by sexually penetrating him, an act that, theoretically, neither of us wants to perform in real life.

HOMOPHOBIA AND THE FUCKER

The paradoxical dynamics of homophobic cussing are made much more intelligible when we realize that homophobia is not synonymous with hatred of homosexuals. In "Masculinity as Homophobia," Michael Kimmel says, "This, then is the great secret of American manhood: We fear other men. Homophobia is a central organizing principle of our cultural definition of manhood. Homophobia is more than the irrational fear of gay men, more than the fear that we may be perceived as gay men . . . Homophobia is the fear that other men will unmask us, emasculate us, reveal to us and the world that we do not measure up, that we are not real men" (2001, 35).

Straight men are taught to deal with their fear of each other by indulging an obsession with violence. As Kimmel (2001) states, "Violence is often the most single evident marker of manhood" (35), even if most men are rarely violent in their day-to-day lives. Obsession with violent sports, video games, movies, music, and speech is a sign of a man's investment in heteronormal manliness. It is also a veiled warning to other men because it implies willingness to engage in violent behavior if provoked. Ultimately, the projection of violence is a means to ensure other men will not violate his masculinity, with rape being the most graphic nightmare scenario.

Such is the motivation behind *fuck* as code for hurt, damage, or ruin. But there is something more basically wrong with fuck than its most extreme expression as men raping men. To fuck is different from making love in that the fucker cares little about the fuckee, while the lover is tender with the beloved. The word *fucker* gets its power as an insult from the contrast between the mindless, soulless fucker and the sensitive, praiseworthy lover. But the lover's sensitivity may be considered a sign of weakness since sensitivity is associated with femininity, softness, and homosexuality. In the same vein, calling a man a fucker can also be a compliment because a fucker is a hard, macho penetrator, not a sensitive, soft girly-man. Such fucker-fuckee language translates from individual to team, a group of men who take on characteristics of intimate corporate identity, that is, they behave as one body-mind as well as individuals. This unified body-mind of the team is *socio-somatic*, a manifestation of group solidarity with shared dress code, behaviors, slogans, and attitudes.

Homophobic speech takes on an entirely different dynamic, however, when it occurs outside of a team's intimacy and enters the realm of detached official discourse. Logic dictates that the fucker has, at least figuratively, a hard-on. This is the unavoidable paradox of homophobic insults: in order to turn another man into a fag, the insulter must be sexually aroused by the man he would turn. Once this is pointed out, there is but a short step to public scandal and the burning ego-scorch of comedy, especially the Gay folk performance of camp. It is crucial (yet impossible) that the fucker and his beloved sports/military teams remain the polar opposite of the effeminate homosexual, "effeminate" and "homosexual" being synonymous.

LET'S ALL BE FAIRIES

Notions of masculinity and effeminacy were fuel for humorous performances of camp in the Gay community during the early twentieth century, and these performances became popular across the United States, eventually making their way to military bases. In contrast to the military/sports heroic ideal, cross-dressing males (pansies or fairies) were antiheroes. But much like another antihero, the fucker, the pansy-fairy was both repellant and attractive. Large urban centers during the 1920s experienced the "pansy craze," and effeminate men became stars in cabaret shows.

The pansy craze generated humorous songs about effeminate men, such as "The King's a Queen at Heart" (about the cross-dressing propensities of England's King Edward, sung by Judd Rees, 1934), "Green Carnation" ("We are the reason for the nineties being gay," Noel Coward, 1933), "I'd Rather Be Spanish than Mannish" (Jean Malin, 1933), and "Let's All Be Fairies" (the Durium Dance Band, 1933, www.queermusicheritage.us). The lyrics of "Let's All Be Fairies" tie together same-sex romance, effeminacy, and nonviolence. ("Dash," in the second verse, refers to the name of an infamous Gay bar in San Francisco that was shut down by authorities in 1909.)

Two great big burly boxers
Were engaged to appear in the ring
Said the slosher to the slugger
I won't hurt you
If you don't hurt me, old thing
When we're in the ring tonight
There's no reason why we should fight

So let's both be fairies
Tinkle tinkle gnesh gnesh gnesh
Don't get annoyed if I shove you
You can't imagine how I love you
We'll punch flimsy-flamsy
You go out when I say "dash"

[chorus] Ding dong ding dong

Fairy bells are gaily ringing
Ding dong ding dong
Everybody's gaily singing

Dancing 'round the moonbeams
La-la-la-la, la-la
Hear our fairy footsteps
Whoops, there we are

The gist of these verses is this: if both big burly boxers become fairies dancing 'round the moonbeams, nobody gets hurt. Since a willingness to be hurt (to "take it") is central to macho identity, a desire to avoid being hurt must be viewed as cowardly or at least antiheroic—a notion that is easily extended from the athlete in the boxing ring to the soldier on the battlefield. Consequently, emasculation of real men by the antiheroic antics of fairies was not simply perceived as shameful or absurd. By the 1930s, fairies and camp were treated as threats to national security. Authorities shut down pansy shows in the big cities after the repeal of Prohibition, and vice squads arrested fairies for corrupting men in uniform during World Wars I and II.

But even then, the official stance did not eliminate pansies in shows produced by warriors for warriors. Military bases staged performances with men in drag to entertain troops during World War II, the assumption being that presumably Straight male soldiers in drag were harmless fun. Leila Rupp writes, "Even extremely effeminate men could play the barracks 'fairy,' a comic role familiar to many from the 1920s 'pansy craze.' All-male (and racially segregated) shows that soldiers put on for the troops gave some men the chance to dress in drag. Army Special Services put out handbooks with scripts, music, lyrics, set designs, and even dress patterns" (1999, 136).

IN THE NAVY: BISHOP TRADING SEAMEN

Lack of clarity concerning homophobia and masculinity caused a scandal in Newport, Rhode Island. In 1919, a covert operation was launched to stop homosexual activity involving servicemen at the Naval Training Station after a sailor reported effeminate behavior, cross-dressing, and illegal intoxicants. Handsome young seamen were ordered to hang out at places such as the local YMCA and entrap men who sought sex with them. The sting was a success until charges were brought against Samuel Neal Kent, an Episcopal bishop. Typically, accused men accepted their punishment quietly so the whole thing would quickly go away. But Kent did not take it lying down, and the bishop's resistance made headlines. During the course of Kent's trial, sailors described in graphic detail their own willing participation in sex acts as evidence of the accused's guilt. Military folk beliefs about masculinity did not consider the penetrator in homoerotic sex (a penetrator was known as *trade*, and finding one for casual sex was *trading*) to be immoral or emasculated. Only the penetrated—known as *perverts, fairies*, and, in this instance, the *Ladies of Newport*—were blameworthy. The public was shocked by the news that military authorities sent out young servicemen to have sex with fairies, and the Navy dropped all charges (Loughery 1998, 3–12).

The Newport scandal highlighted the need for official silence concerning homosexuality. The absurdity of military men's folk classification of the penetrated male as the only homosexual in the act was laid bare in a court of law. Any homosexual behavior, regardless of who did what to whom, was indicative of homosexual identity, regardless of how masculine or unpenetrated a participant might be. But old ways die hard, and effeminacy would continue to be the defining mark of male homosexuality. Fifty years after Newport, the Stonewall Rebellion (1969) in Greenwich Village hit the presses. Photos of young Gay men fighting back against police shocked the world. Pictures notwithstanding, the three-day insurrection was typically portrayed as a team of limp-wristed, dancing pansies against brutal macho police officers, which in its own way was even more embarrassing for authorities. Queer men never hit Straight cops so severely as when they resorted to violence while defiantly prancing about like girls. The Stonewall episode vividly undermined the fiction of fearless heteronormal male rape (and consequential effeminate weakness and passivity of the raped).

But the Gay male community was by no means limited to iconic dancing queens. Men were busy creating their own macho identity in the streets and dance floors of Manhattan, a movement that launched the disco craze of the 1970s. Sixty years after the Newport scandal broke, the clearly uneffeminate men in the Village People released "In the Navy" (1979), a dance hit with playfully homoerotic lyrics. The physically fit bodies, masculine bearing, assertive dance moves, and manly voices of the Village People fooled the United States Navy into allowing the group to shoot a video aboard an American warship in exchange for rights to the song. Plans for using "In the Navy" in recruiting commercials sponsored by the Department of Defense were abruptly cancelled when officials discovered the reputation of the Village People as non-*nelly* (effeminate) but suspiciously Gay (Prono 2008, 272).

BATTEN DOWN THE HATCHES

Dance continued to play an important role in redefining Gay men's masculinity beyond the pansy stereotype. Circuit parties, weekend-long events where men gather to dance, get inebriated, and flirt, began in Manhattan soon after Stonewall, spread nationwide during the late 1980s, and went international in the 1990s. Most Circuit boys do not conform to the image of the Gay man as effeminate. They do not speak or dress in an effeminate manner, and more than a few are physically fit. For most of its history, the Circuit was ignored by the American public, possibly because the sight of so many muscular men dancing together did not conform to the cherished pansy antihero stereotype. Many Circuit weekends include military balls where participants dance in military clothing.

This relative invisibility changed in May of 1996 when Cherry Jubilee weekend debuted in Washington, DC, and was brought to the attention of the US Congress by California representative Robert Dornan. Dornan was, in his own words, "a God-fearing American, a very lucky husband of 41 years, a father of 5 stalwart, God-loving adult children, a grandfather of 10—No. 11 is in the hanger" (*Congressional Record,* June 27, 1996). On May 9 he spoke before the House of Representatives about Cherry's use of a government building for a brunch, condemning the event as an out-of-control orgy of open-mouth kissing, men in drag, simulated sex, and drugs (Weems 2008 124–129). His presentation led to bickering between conservatives and liberals concerning the homosexual agenda, Gay-bashing, and AIDS in America. On June 27, he again harangued his colleagues about the evils of Cherry Jubilee. Supporting his own position with references to the Holy Spirit, Moses, the Reverend Billy Graham (and his wife, Ruth), Pope John Paul II, Holy Mother Church, and Christian love, he contrasted the moral depravity of homosexual men with the discipline of real men in uniform: "Imagine for a moment, Mr. Speaker, if the out-of-control homosexual romp that we judge today had happened on any US military base or post anywhere throughout the world. What would the repercussions have been? Batten down the hatches . . . how dare we live by a lower, a much lower, standard of ethics and professionalism than we demand of our younger military men and women who serve under our jurisdiction, and who do risk their very lives . . . our Congress ignores garbage like this 'Cherry romp' of hedonism right here down on Constitution Avenue" (*Congressional Record,* June 27, 1996).

Dornan also implied that Cherry Jubilee was worse than the 1991 Tailhook scandal in Las Vegas, when ninety people (most of them women) were physically assaulted and sexually harassed by men attending the thirty-fifth Tailhook Association Symposium, an annual event for Navy and Marine aviators. Try as he may, Dornan's efforts to generate outrage failed because Straight Americans (homophobic or otherwise) were not comfortable comparing military men to hot, muscular men under the disco ball.

FROM FAG BOMB TO GAY BOMB: ADVERSELY AFFECTING ENEMY UNITS

If Dornan was correct in his belief that horny, drug-addled, shirtless, dancing Gay men undermine the integrity of the US Armed Forces, it stands to reason that an increase in homosexual activity behind enemy lines would debilitate said enemy. Such was the logic informing the "Gay Bomb," a weapon that was supposed to cause irresistible man-on-man lust and that was the subject of research at Wright-Patterson Air Force Base, which had been given $7.5 million for nonlethal chemical weapons research. A 1994 report entitled "Harassing, Annoying and 'Bad Guy' Identifying Chemicals" describes a bomb that would release an aphrodisiac, causing the spit-and-polish heterosexual discipline of the bad guys to fail (Petras and Petras 2007, 259–260). This would result in what Dornan would no doubt consider to be a massive "Cherry romp" gone ballistic, a sexual free-for-all in which "discipline and morale in enemy units" were "adversely affected" by a feverish epidemic of open-mouth kissing, cock sucking, butt fucking, and booty shaking. References to the Gay Bomb as a viable weapon of ass destruction appeared again in 2000 and 2001 (Hambling 2007, 25).

The underlying notion behind the Gay Bomb is that, by generating rampant sex between men, the weapon would transform the enemy into the flimsy-flamsy pansy stereotype expressed in "Let's All Be Fairies." That such an idea could even be considered demonstrates that fears and fantasies of military homophobic discourse continued to titillate important people in uniform in the twenty-first century. Once the project was made public, the absurdity of the Gay Bomb earned an Ig Nobel Prize for the Air Force in 2004. No representative from the US military accepted the award, a trophy featuring a chicken attempting to swallow an egg.

PINK SWINGING DICKS

Those same homophobic-erotic fears and fantasies that fed scandals can be found in enlisted men's folk display. Transgressive folk speech concerning homosexuality was part of my boot camp experience when I joined the Marine Corps in 1983.

Enlisted military personnel are not trained by commissioned officers. Since the Revolutionary War, the American military has followed British protocol and placed the training of common soldiers in the hands of sergeants rather than officers, with the outstanding exception of General Friedrich von Steuben, the young nation's first official drillmaster. During the American Revolution, General George Washington turned a blind eye toward Steuben's homosexual tendencies while dishonorably discharging other men for the same offense. Ironically, perhaps the earliest account of humorous obscene language in American military training comes from descriptions of the closeted homosexual general drilling his poorly disciplined troops. According to Roland Young (1978), "When troops misunderstood or performed badly at drill, he [Steuben] would lose his temper . . . and swear wildly in German and then in French, mixing in a few 'God-damns,' one of the few English phrases he knew. When he had exhausted his supply of oaths, he would call one of his aides: 'Come and swear for me in English . . .' The soldiers loved it. Drilling under Steuben was high comedy" (71). We do not know if Steuben's colorful language included homophobic-erotic language. But in light of his orientation, his performance certainly qualifies as camp.

Trained separately from female Marines, male Marines are exposed to homophobic-erotic speech while they are in the amorphous state known as "recruits" and under the constant supervision of professionals whose job it is to mold them into "mean green fighting machines." (Green is often used to represent Marines as a siblinghood that transcends race. There are no Blacks or Whites in the Corps, only dark green and light green.) Under the jurisdiction of DIs, USMC boot camp is transformative. When one becomes a Marine, one is a Marine forever. Recruits are "Born Again Hard," calling to mind the death-liminality-rebirth pattern found in boy-to-man rituals. For male recruits, USMC boot camp is the place of *androgenesis*, where men give birth to men.

Although drill instructors may not describe their work in those terms, they do see themselves as the agents of rebirth who transform a recruit from an undisciplined and *unsat* (unsatisfactory) thing into a higher stage of disciplined existence, typified by the complimentary phrase "locked and cocked" (as in a weapon ready to fire, signifying a good Marine). They do this by means of close daily interactions with recruits as they break them down, reshape them, and build them back up as individuals with a new socio-somatic corporate identity. The relationship between recruit and instructor is thus an intimate one, fostered by performances carefully choreographed to effect the transformation. Never are recruits left alone to fend for themselves. Drill instructors get them up every morning, put them to bed every night, and sleep over with them for three months in the secluded world of their squad bay, one of many such crèches in the Recruit Depot.

Boot camp is an idealized world with two performance frames: with and without officers. As mentioned earlier, drill instructors are never officers. They are in the lower ranks and pay scale of enlisted personnel. Symbols for pay grade are O (officer) and E (enlisted). In the intimate boot camp frame of without-officers, my senior drill instructor referred to officers as "zeros," drawing a distinction between the higher and lower classes of military hierarchy in favor of the lower. Never would he use such a term to a zero's face.

Get enlisted Marines together, and a favorite topic is the antics of our drill instructors, those consummate performers who were the living embodiment of what a Marine should be. The best stories are situated in the without-officers frame. Since homophobic-erotic language is officially forbidden (as is any form of cussing), it was reserved for the intimacy of the squad bay when drill instructors had us all to themselves. Their words followed us when we left the Island (the Marine Corps Recruit Depot on Parris Island, South Carolina) or Hollywood (the recruit depot in San Diego) and entered into our own folk speech. Humor is its most important element.

Although the Marine Corps neither officially recognizes nor endorses homophobic-erotic folk speech, Hollywood (in LA) has spread it far beyond recruit depots, bases, and bars. Scriptwriters inserted homophobic-erotic language in the 1987 Warner Brothers film *Full Metal Jacket*; in its boot camp scenes, the drill instructor says the following:

God has a hard-on for Marines.
  I bet you're the kind of guy that would fuck a person in the ass and not even have the goddamn common courtesy to give him a reach-around.
  Do you suck dicks?
  I bet you could suck a golf ball through a garden hose.

Movies such as these inspire many would-be recruits to expect the very same language when they go to boot camp, as I did from movies I had seen before my time in Parris Island. But phrases come and go, as do regulations concerning proper language in the squad bay. Outside of "assholes to belly buttons" (men jammed together in close quarters, also called "nuts to butts") and "pink swinging dicks," I rarely heard my drill instructors cuss. Instead, they would use "doggone" instead of "goddamn," "trash" instead of "shit" ("I'm tired of your trash, recruit!"), and refer to us as "thing" instead of "son of a bitch" or any other cuss word. Nevertheless, my drill instructors would occasionally use homophobic-erotic language, such as asking a recruit, "Do you want to fuck me?" if the recruit looked the drill instructor in the eye. When one recruit requested an "emergency head call" (he needed to use the bathroom) and could not stand still in formation because he had to urinate, one of my drill instructors told him that he was giving the drill instructor a hard-on. "My wife moves like that when we have sex!" the drill instructor added for our amusement. At that late point in our training, we were permitted to laugh.

RIFLES, GUNS, AND THE BIG GREEN WEENIE

Sex-as-violence is present in homophobic-erotic boot camp speech, but as a minor theme rather than a dominant one, which makes it markedly different from sexual penetration expressed in men's insults. Phallic resonance between the penis and the rifle expressed in "locked and cocked" also links the penis to a weapon of death rather than an instrument of pleasure and procreation. But the link is a humorous one, as in the following rhyme taught to recruits and passed on to the civilian world:

This is my rifle, this is my gun [penis]
This is for fighting, this is for fun.

An example of phallic metaphor is the "Big Green Weenie up your asshole" in Marine Corps folk speech. The Big Green Weenie is USMC bureaucracy portrayed as an uncaring, mindless schlong that rapes helpless Marines and can "take a dump" (defecate) on you. Being figuratively penetrated or shat upon by the Big Green Weenie denotes a shared experience of abuse inflicted on soldiers by a powerful force that cannot be resisted.

The humor implicit in the image of a massive green penis (I envision it as a green horizontal cartoon hot dog) lies in its position as a threatening yet beloved entity, not unlike a drill instructor. But not identical: Marines tend to look back on their drill instructors as sadistic but concerned parents, exquisitely brutal caretakers who ultimately have recruits' best interests at heart. The difference between the idealized world of Parris Island/ San Diego and the gritty world of the everyday Corps is often a harsh one. Drill instructors can indeed be real pricks, but at least there is a person attached to the appendage. USMC bureaucracy, on the other hand, is a faceless dick, dismembered, discolored, and detached from the problems of everyday Marines.

Nevertheless, the purpose of the Big Green Weenie in enlisted Marine folk speech is to signal solidarity with other enlisted Marines, often framed as "real Marines" as opposed to pencil-pushing zeros. Affection for the Corps is ultimately grounded in the enlisted ranks more than in the officer, which is why some USMC officers go through enlisted boot camp before going to Officer Candidate School.

FUCK-FUCK GAMES

Except for funerals, the love that male military personnel have for each other has limited expression in official military discourse. "Love" is mentioned in the Marine Corps in three ways: love of God, Country, and Corps, in that order. It is given indirect reference in *esprit de corps* ("spirit of the body," the socio-somatic identity of a group as one being). But in the regimented bodies and acronym-studded cant of everyday *officialese*, or official language of the military (see Levy's chapter 5 in this volume), there is little room for tenderness among men who take seriously the work of defending the nation.

In private folk speech and performance grounded in the *enlistic*, folk speech of enlisted personnel (see Levy in this volume), love between men is often couched in homoerotic humor, which may include affection for their closeted homosexual fellows. This could be especially true for sailors in the close confines of life at sea. My father remembers that when he was in the Navy during World War II, everyone knew who on ship was homosexual and who was not. Those who were homosexual were called "the daisy chain," referring to a sexual position of men in a circle, mouth to crotch to mouth to crotch, and so on. The term was used in the spirit of teasing rather than malice.

Whether at sea or on land, it is useful to think of enlisted life as a form of incarceration with benefits, and of officers as wardens. This is reflected in our language: people "serve time" in both the Armed Forces and penitentiaries. When we grant our ground-pounders, artillery, submarine crews, and flight deck personnel access to powerful weapons, we keep them on a very short leash. This goes to the root of what it means to be civilized. If the purpose of civil law is to prevent vendetta, then war is the failure of civil law. Military personnel are not civilians, that is, they are not bound by civil law when on the battlefield. Warriors in civilized society are licensed outlaws, deserving of incarceration as long as they have access to deadly force. Their bodies are conditioned to instant obedience through drill, rendered immobile on demand when standing at attention, and decorated with special markers on splendid uniforms that are identical to those of their fellows, much like glorified prison garb. Their quarters are open to scrutiny at all times. They cannot quit if they do not like their job, neither can they strike for better working conditions.

But constant supervision of enlisted personnel comes with a price. Military personnel are expected to behave with unrelenting discipline and decorum, and enlisted men find ways of venting through transgressive masculine behavior befitting an outlaw identity just beneath the veneer. Sexually explicit or repugnantly violent marching songs are one means. Death-related images such as skulls, serpents, blood, swords, and flames tattooed on the body are another means. There are also rituals involving pain, endurance, and homophobic-erotic performance.

Officers are not immune to the need for transgression, but it is increasingly forbidden to them since they represent the eyes and ears of the institution. If transgression is kept in the realm of the unofficial, untranscribed, and plausibly undetectable (such as homophobic-erotic language in USMC boot camp), officers can rest easy. And if it can be framed as play, all the better. Mildly sadomasochistic rituals situated in a playful, humorous frame have served as release valves for the pressures of relentless discipline, such as pounding jump-wing pins into bare skin (blood wings) by Army and Marine paratroopers. Such rituals as blood wings and centuries-old traditions such as Navy shellback initiation at sea are, however, being erased or driven underground because they have crossed the line from team business to public scrutiny (Bronner 2006, 47–54), which in turn forces officers to clamp down even harder on enlisted personnel.

Conflict between zeros and enlisted is unavoidable. Callous abuse is purposefully built into Marine Corps boot camp training to prepare recruits for the disconnect between themselves and those in command. My drill instructors called it "playing games." If we did not react fast enough, precisely enough, or gung-ho enough to satisfy a drill instructor (or even if we did), he would announce to us, "So we want to play games, do we? Okay, we'll play games!" This was the preamble to a temporary world of shit where we would be forced to do painful physical exercise, mind-numbing repetitive drills, or on-our-knees deck scrubbing. The purpose of these games was to make us suffer together, obey without question, and prepare us for inevitable clashes with superiors.

In the greater Corps outside of boot camp, such shenanigans are called *fuck-fuck games*, orders that generate unnecessary stress and often lack fairness, such as a punishment imposed by a higher-ranking person on lower-ranking personnel because of a perceived slight or unsatisfactory performance. Fuck-fuck games abuse the incarcerated body by making men go on *butt patrol* (picking up litter), do extensive PT (physical training), clean the barracks if one failed inspection (or if one didn't, but the superior had a grudge), and remain confined to quarters when off duty.

But fuck-fuck games are not always top down. They can also be bottom up, as dramatized in Anthony Swofford's memoir *Jarhead* (and the film based on it), when enlisted Marines stage a "field fuck," a simulated man-on-man orgy, in front of embedded civilian media in retaliation for mistreatment by their superior. Simulated same-sex orgy is a men's folk tradition of festive transgression and comic relief that occurs in sports teams as well as military teams. Enlisted men having a field fuck in the face of the institution's eyes and ears purposely cross the line. Although the only references I can find to the actual term *field fuck* come from *Jarhead* itself (Swofford 2003, 20–22), male Marines with whom I have spoken are familiar with it and have themselves witnessed simulated orgies. One source actually referred to the orgy as a fuck-fuck game, whether done in the privacy of the team or not. This makes sense when we consider the utterly transgressive nature of a field fuck. It is an unspoken team challenge to military authority over men's bodies that, before the fall of DADT in 2011, officially did not tolerate even the mention of homosexual identity and performance.

As Americans became more Gay tolerant, the institutionalized homophobia of DADT was increasingly seen as a top-down fuck-fuck game, an added stress on the team based on what some men's bodies were not allowed to do in private, not even in the virtual world of the Internet. In the intimate sphere of the squad bay, barracks, and hangar, men learn all kinds of things about each other, including sexual orientation. If a trustworthy member of one's unit is afraid of being singled out and punished for what he does as a sexual being, it affects those around him. DADT undermined the vital unity necessary for the team to function efficiently and, above all, survive under fire. Institutionalized homophobia created further resentment against the bureaucracy, not only among enlisted but also in officers who did not want to lose valued members under their care.

MILITARY INTELLIGENCE

A new folk art form is available to members of the armed forces: the humorous video posted on YouTube that features a fuck-fuck game. Some games of festive transgression involve tests of pain, such as men kicking each other in body armor, setting themselves briefly on fire, or playing with Tasers in drive-stun mode (a setting for a Taser in which the cartridge for shooting barbed electrodes is removed, producing a visible electric arc that can be inflicted upon anyone within arm's reach). It is not unusual for Taser videos to feature shocks delivered to bare buttocks or chest, and there is at least one video of a soldier Tasering his testicles (pants on). Part of a typical performance in the test of pain is the grimacing, shrieking, and laughing as each man takes his turn. What may seem only tomfoolery among guys with too much time on their hands is also the conscious decision to use military property (the body of a warrior) in ways the team, not the bureaucracy, chooses. Within their private sphere, the team makes its own grab-ass rules.

Even more transgressive are some videos in which male soldiers dance together. Often the dance moves are sexually suggestive. Even more scandalous are videos of men dancing while on patrol, or in full battle gear and wielding firepower, using their weaponry as props in their choreography. Other videos show US military men dancing with male Iraqi/Afghan military personnel and civilians. American men dancing with each other like female strippers, bent over with butts waving around (sometimes grinding in the crotch of a buddy), is understood as comic from the American cultural perspective. But when such moves are done in full combat gear while dancing with Iraqi or Afghan men (who typically dance with each other in public settings rather than women because of cultural-religious restrictions), they challenge local sensibilities concerning gender and sex as well as potentially offending military superiors.

Throughout this essay, the private world of enlisted men has been contrasted with the official world of officers and the bureaucracy of the military-as-institution. This is in keeping with a definition of folklore as the study of any group bonded by investment in a shared intimate identity (a folk) that generates its own aesthetic and ethical expression (folklore and folklife) outside of bureaucracies regulating that group. We may look at the armed forces in two ways: military-as-institution (a group that imposes impersonal regulation) and military-as-folk (a group that makes its own rules, in this case the folk being enlisted men), and examine the interplay between the bureaucratic and the intimate identities of the folk. We may also see examples of folklore with significant depth in aesthetic and ethical creation, revealing intelligence in design expressed within the context of that group's folklife.

Intelligent design can be found in aesthetic and ethical expressions of a folk, any folk in fact, in those displays that reflect deeper significance and sophistication, all the while using the language, material culture, and performance codes favored by that folk. Such sophistication from enlisted-men-as-folk is evident in *Dance Party in Iraq*, a video that juxtaposes the song "Electric Avenue" by Eddy Grant (1983) to footage of soldiers in Iraq. Alongside the somber is the hilarious. Humorous homoerotic scenes involving a man in an outdoor shower lip-synching a disco song, a man in the latrine, and men dancing together (often with sensuous dance moves) are mingled with scenes of a tattered Iraqi flag fluttering in the wind, Iraqi captives, children running behind patrol vehicles, armed personnel in helicopters, a camel, distant explosions, and general grab-ass behavior (including two friendly dance scenes with civilian and military Iraqi men) strung together in a painfully poetic narrative, all of it unspoken.

Videos cast to the virtual winds of the Internet can create problems for those caught in the act, whatever that act might be. Body talk, especially dance, is a favored medium of speech, perhaps because recorded spoken word, even in jest, can lead to disciplinary action. This was the case of humorous videos produced on board the USS *Enterprise* in 2006–2007 by Captain Owen Honors. Men showering together, women showering together, men sleeping together, references to masturbation, and a cameo by actress Glenn Close are almost incidental to the scandal, which primarily focused on Captain Honors's use of "fag" and "gay." But when seen from the perspective of the team-as-crew and in the context of Honors's performance of multiple roles, "fag" and "gay" were not necessarily slurs against the LGBTQ community. The role Honors plays when he uses those words is that of an arrogant, supermacho aviator of the Tailhook scandal stereotype, the fucker that nobody likes. That character's use of the terms is less an attack on Gay sailors than a sly indictment of homophobic intolerance.

A notable exception to the unspoken rule concerning homosexuality is *Field Day Linebacker–USMC*, a video about Big Sexy, a "barracks linebacker" who physically tackles men who do not follow the rules, especially on field days (barracks cleaning days). In one scene, Big Sexy tackles two stereotypically effeminate male soldiers for ignoring DADT. Even then, not once does Big Sexy use "faggot" or any other orientation-related slur. Later in the video, it is revealed that supermacho Big Sexy has his own dalliances with other men.

AND I'M TELLING YOU

In 2011, the US military dismantled Don't Ask, Don't Tell and began admitting openly Gay male and Lesbian personnel into its ranks (transpeople, however, are still excluded). The reaction of most Americans, including members of the armed forces, has been neutral to favorable. This tolerance is expressed in a video attributed to Codey Wilson called *If the Army Goes Gay: A Cody Wilson Production* (Dont Ask Dont Tell 2011), a spoof of military life after the repeal of DADT. There is a brief dialogue at the beginning:

What?

What's wrong?

Gays, man! Gays in the military, it's alright to be Gay in the military!

Participants in this folk video flout the rules against homosexual behavior in several culturally encoded ways, most involving dance, as if Dornan's feared "Cherry romp" had actually come true. There are scenes of shirtless men dancing on top of vehicles, men watching each other in the showers and at urinals, men sleeping together, giving each other massages, accessorizing military gear in very unsat outfits (such as body armor and nothing else), dancing together in choreographed routines (one scene with glow sticks), and a stylized four-man field fuck. At the end of the video, a picture of a stuffed toy wolf being sodomized by a stuffed toy Pikachu is the background for the following statement posted "For protection purposes": "This feature was 100% my idea. My great step team CREW DAWN deserves the respect. Should you be offended no one else is to blame. Besides you don't really want to cause a few young brave soldiers trying to have a good time any more stress in a combat zone do you? No. Just be honest with yourself. You loved it."

Although the performance is obvious buffoonery, the unspoken message for both superiors and civilians is this: we will act as flimsy-flamsy queer as we wish, as macho-queer as we wish, dancing in the moonbeams whenever and wherever we wish. As such, videos such as this one undermine distinctions between the masculine Straight warrior-hero and nelly homosexual antihero in favor of team solidarity and mutual affection that transcend both homophobia and homoeroticism. Within the transgressive world of military humor where nobody is above being teased, these videos make room for openly Gay comrades as full-fledged, beloved members of the team.

WORKS CITED

Bronner, Simon. 2006. *Crossing the Line: Violence, Play, and Drama in Naval Equator Traditions.* Amsterdam: Amsterdam University. http://dx.doi.org/10.5117/9789053569146.

Hambling, David. 2007. "Gay Bomb." *New Scientist* 96 (25): 2624–29.

Kimmel, Michael S. 2001. "Masculinity as Homophobia: Fear, Shame, and Silence in the Construction of Gender Identity." In *Men and Masculinity: A Text Reader,* ed. Theodore F. Cohen, 20–41. Belmont, CA: Wadsworth Thomson Learning.

Loughery, John. 1998. *The Other Side of Silence: Men's Lives and Gay Identities; a Twentieth-Century History.* New York: Henry Holt.

Petras, Kathryn, and Ross Petras. 2007. *Unusually Stupid Politicians: Washington's Weak in Review.* New York: Villard.

Prono, Luca. 2008. *Encyclopedia of Gay and Lesbian Culture.* Westport, CT: Greenwood.

Rottman, Gordon L. 2007. *FUBAR: Soldier Slang of World War II.* New York: Osprey.

Rupp, Leila J. 1999. *A Desired Past: A Short History of Same-Sex Love in America.* Chicago: University of Chicago Press.

Swofford, Anthony. 2003. *Jarhead: A Marine's Chronicle of the Gulf War and Other Battles.* New York: Scribner.

Weems, Mickey. 2008. *The Fierce Tribe: Masculine Identity and Performance in the Circuit.* Logan: Utah State University Press.

Wright, John D. 2001. *The Language of the Civil War.* Westport, CT: Onyx.

Young, Rowland L. 1978. "What Is to Become of the Army This Winter?" *American Bar Association Journal* 64 (January): 68–71.

YouTube VIDEOGRAPHY

"Dance Party in Iraq." 6,499,688 views as of July 17, 2011. www.youtube.com/watch?v=UW1toLy_FMQ.

"DONT ASK! DONT TELL . . . soldiers in iraq doin the blah blaH Blah." 4,280 views as of July 17, 2011. www.youtube.com/watch?v=ciPT-4FvsKI&feature=related.

"Field Day Linebacker—USMC." 51,318 views as of July 19, 2011. www.youtube.com/watch?v=ONVDU0icLYc.

"Fun With Taser Pt. 2" 13,164 views as of July 17, 2011. www.youtube.com/watch?v=J8X4kLM4GEA&NR=1.

"Infantry bored in Iraq." 9,998 views as of July 17, 2011. www.youtube.com/watch?v=cxm gQzrC0is&feature=related.

"Taser Fun." 5,225 views as of July 17, 2011. www.youtube.com/watch?v=R3PryFqUNyo&NR=1.

"USS Enterprise Video Scandal with XO." 291,401 views as of July 17, 2011. www.youtube.com/watch?v=srbLyuMgDe8.

"US Soldier gets bored and tasers his nads." 25,104 views as of July 17, 2011. www.youtube.com/watch?v=rd20GIkERHE.

"When Marines get bored in 29 Palms." 434,277 views as of July 17, 2011. www.youtube.com/watch?v=uXANHKFECbI&NR=1&feature=fvwp.

8

Making Lemonade
Military Spouses' Worldview as a Coping Mechanism

Kristi Young

Oh, spouses are in the military! . . . You're marrying an institution here.

—Canadian general (quoted in Harrison and LaLiberte, 1997)

Military life is what you make it.

—Lindsay Madsen, wife of a US Air Force officer

In *Wives and Warriors: Women and the Military in the United States and Canada*, editor Laurie Weinstein writes that when their spouses are present, military wives "must ascribe to the norms of femininity (be passive, submissive, and dependent on men), yet when the men go on deployment, these same wives are expected to be leaders and decision makers" (Weinstein and White 1997, xvii). The psychological conflict inherent in this situation is a significant factor in what she and coeditor Christie White see as many military wives' negativity about the armed services—rigidly structured institutions that rely implicitly on spouses' unpaid labor and that see spouses themselves as "shadow enlistees" who must function as "perpetually accommodating" multitaskers (Harrison and LaLiberte 1997, 38–39). Laurie Weinstein shares an anecdote from her own experience: "Years ago I attended what would be one of my first of over fifty Navy parties. As a newly engaged bride-to-be, I was eager to meet everyone and make a good impression. As our engagement was announced to one captain's wife, she looked at another officer's wife and said, 'Should we tell her?' Ten years later, I know

DOI: 10.7330/9780874219043.c08   161

exactly what she meant. Her pointed question could not belie the months of sadness, loneliness, frustration, and anger she must have endured during her many years as an officer's wife" (Weinstein and White 1997, iii).

There is no question that, for many wives, being "married to the military" can have negative aspects. Military life can be strenuous and dangerous, and the work hours can be horrendous, even when members on active duty are not deployed. In addition, while service members may enlist enthusiastically for a variety of reasons, most of the women married to them do not grow up dreaming of being military wives. Nor are they always proactively involved in choosing this way of life; often it is thrust upon them as a consequence of marriage. So military spouses, forced to adapt to a regimented lifestyle not of their own choosing, and periodically physically separated from their husbands, may be expected to feel sometimes constrained, resentful, and lonely.

Indeed, it is because such feelings are common that the services have created family support groups to prevent spouses from floundering unhappily in isolation. To build what social scientists call "community capacity," they have also encouraged networks of informal friendship groups, where women whose men are deployed can gather for coffee, games, playtime with their children, and (perhaps most important of all) conversation. "It is these informal group associations," according to one study, "that are most accessible to individuals and families and those whom they most often rely upon on a daily basis" (Huebner et al. 2009). It is also in these informal groups—folk groups that are of but not officially in the military—that military wives are most readily able to express the negativity noted by Weinstein and White.

But griping about the constraints of their lives is not the whole story. I have been studying a group of these "shadow enlistees" for some time now, and I have found consistently that while negativity is certainly an element of some of their conversations, they also express pride and other positive feelings about the services; reinforce the rules of behavior required of them; find humor in their shared situation; and employ anecdotes and longer narratives as coping mechanisms. Such positivity is seldom acknowledged in critiques such as Weinstein and White's, yet it is a real aspect of these women's lives, and a defining element of their shared expressive behavior. I focus in this chapter on this positive aspect, particularly on the role of the women's worldview in proving the maxim—variations of which I heard often in my interviews— "Military life is what you make it."

I say worldview rather than "worldviews" for two reasons. First, the wives that I have interviewed are nearly all members of a discrete subset of military spouses: they are members of the Church of Jesus Christ of Latter-day Saints, and most are married to men in the US Air Force. I suspect that their views of military life are similar to those of women who are not Mormons and whose husbands belong to different branches of service, but I have no direct evidence of that; so I offer this essay as suggestive rather than definitive.

The second reason for speaking about a singular worldview follows from the first. While the military spouses I know differ widely in their interests and personalities, they share a common set of what Alan Dundes called "folk ideas," that is, the "traditional notions that a group of people have about the nature of man, of the world, and of man's life in the world" (Dundes 1971). Among the military wives I have interviewed, such traditional notions include a deep respect for resourcefulness, for personal friendships, for duty to country, and for religious faith. They speak about these units of worldview, often deliberately and consciously, as elements of a coping strategy that helps them weather the strains of military life.

I first became interested in military wives several years ago, when my daughter Lindsay married an Air Force missileer, John Madsen; they have been stationed for the past two years at Malmstrom Air Force Base in Great Falls, Montana. The stories she told me about her life initially sparked my interest in this folk group, and through her I gradually befriended and interviewed other Air Force wives. Most of the stories I discuss in this essay came from those women; others came from friends and associates of my research assistant, Kimberly Keck, whose father and father-in-law were both members of the US Army.

The problems military wives encounter range from the relatively trivial challenges of having to adapt to the intricacies of military protocol to the life-altering shock of a husband lost in combat. When Weinstein and White speak about "sadness, loneliness, frustration, and anger," they are likely referring to that wide mid-range of challenges that are neither trivial nor traumatic but that set the tone, in a sense, of the military spouse's daily life. It is those everyday challenges that I address in this essay as well, and I choose to focus on them for an obvious reason: they are the "lemons" that the women I spoke to mentioned most frequently as they described what it was like to be married to the service. I show here how they turn those lemons into lemonade.

DISLOCATION AND ADJUSTMENT

Among the most commonly noted challenges that my interviewees mentioned was the need to adjust to their husbands' frequent reassignments. This is a nearly universal source of difficulty for military wives. My daughter Lindsay, whose husband John has been in the Air Force only three years, has already lived on bases in Alabama, California, and Montana. Wives of men who have served much longer than that often follow their husbands on even more bewildering sequences of reassignments. Sarah Taylor-Jenson, for example, is the wife of a Marine major who worked his way up from the enlisted ranks. In their twenty years in the service, they have lived in six states, never for more than four years at a stretch. Or consider the even more itinerant journey of Keri Hatch, whose husband, Richard, is an Army attorney:

> The first place that we went was Fort Lewis, Washington. And we were only there for two years . . . And then we moved back east and he worked in the Pentagon. And so we were in the Washington, DC, area—he was at the Pentagon for almost four years. And then we went to Charlottesville, Virginia, where there is a JAG [judge advocate general] school, so they give training to JAG officers, and he earned his master's there. And then he taught on the faculty. It's connected to UVA—University of Virginia's law school. And so we lived there for four years, and then we went to Germany, actually southern Bavaria in Germany. So we were there for two years, then we came back to the States and went to Fort Leavenworth, Kansas. He went to a school there for a year, and then from there we came back to Washington DC . . . for four years. Then we went to Alaska, to Fort Richardson for two years, and then went to Fort Campbell, Kentucky, for two years and then back to Washington DC, where he retired after two deployments to Iraq.

Moving with such dizzying frequency can be difficult—and it is an extremely common experience among the women I spoke to. But most of them, out of necessity, have learned to cope. Their success in doing so, I suggest, has a great deal to do with the folk idea that—to use the old American adage—when life gives you lemons, you should make lemonade. There's a moral imperative implicit in that idea, and a respect for self-reliance and resourcefulness. This idea is deeply embedded in these women's worldview, such that moves that might look overwhelming to others are seen by them as opportunities not just to adapt but to actually enjoy the experience by coming up with creative responses to the dislocations.

For example, Debbie Slik and her husband, Mark, a major in the Air Force, turned one of their reassignments into a fun road trip. As members

of the Church of Jesus Christ of Latter-day Saints, they made the move
a church historical tour in reverse, retracing the route that the original
Mormon families had taken in the 1800s from upstate New York to the
Utah flats:

> Well, we always try to make a big road trip out of it when we move. And,
> like, for instance, we moved from California to Delaware and that was a
> cross-country trip, so we did the pioneer trek in reverse. Went to Salt Lake,
> then to Winter Quarters and Nauvoo, and then Kirtland and kind of snaked
> our way east. So we try to travel and learn while we travel and make a fun
> time of our moves each time. Also, we always buy the kids new sets of sheets
> for their bedrooms so that they feel like, you know, there's something exciting
> to look forward to.

Finding ways to create that "something exciting" is a common theme when
these women discuss their moves. Often, in describing a service-mandated
relocation, they use the word *adventure*. It's a good word, implying not joyful smooth sailing but something more challenging and therefore exhilarating. Tamara Johnson's description of some of her family's relocations is
typical of many:

> Camden [her daughter] was four weeks old when I had packers at the house.
> And we moved and were in the location for three months, and then moved
> across the country all the way over from Virginia to Hawaii and she was
> about five months when we did that move. For most people that would just
> be overwhelming, to have somebody come in and pack the house up with a
> four-week-old baby, and to then drive from Alabama to Virginia and there
> for three month in temporary housing, none of your own stuff. And then to
> pack the few things you do have and load it in the car and then drive cross-
> country, and then fly transpacific to Hawaii and live in a hotel for nine weeks,
> you know, until you get housing. But that's part of the adventure and the
> experiences. And you can either whine and complain and be miserable or you
> can [adjust to it] . . . A lot is the perception and how people deal with things.

Obviously, for Johnson and other wives like her, a positive attitude—the
right "perception"—can go a long way in transforming an arduous experience into an enjoyable "adventure." The wives we've spoken to, almost
without exception, display that attitude. They don't deny that moving can
be difficult, but they are committed to making its very difficulty a source
of enrichment.

We are often told that moving can be especially hard on children, who
are reluctant to leave friends and familiar circumstances behind. Given this,

you might suppose that "service brats" suffer inordinately because of frequent moves. Yet few of the women I spoke to confirmed this supposition. On the contrary, several of them mentioned that, for their children, being forced to move around was a positive learning experience. Keri Hatch, who raised five children while an Army wife, sums it up this way:

> I think they got a worldview that they wouldn't have gotten if we'd stayed in one place. I think they learned how to get along. There's probably not a more diverse population than the military because people are stationed all over and then they end up marrying people from different places. So, you get a pretty wide variety of American society, and also in our situation there were a lot of Korean spouses and some German spouses . . . I would say that my kids would say that they learned a lot and that they got to travel and experience a lot of things that we would never have been able to afford with a family our size . . . They will also say that, for instance, when they went on church missions and went to college that the adjustment was easier because they were used to moving around and making new friends.

Sarah Taylor-Jenson, the mother of three, agrees, making the point that it's the very transiency—and unpredictability—of military postings that stimulates children to develop friendships quickly. In an excellent illustration of the "lemons into lemonade" principle, she points out: "We move quite a bit so you sort of have to get grounded, find a group of people that you connect with and become friends quickly. That's the benefit, my kids know that they usually don't have a long time to be somewhere so they've got to get in there and make friends. That's a life skill that you can't teach otherwise."

When Keri and Sarah speak about the value of making friends, they are referring not just to a valuable life skill for their children, but to a core value in their own adaptive worldview. Indeed, it is the opportunity to make new friends, often just as quickly as their children do, that many of them see as an incalculable benefit of the military lifestyle. My daughter Lindsay speaks to this ironic element when she describes a rapid bond that developed between her and another wife of an Air Force missileer:

> I've formed a lot of relationships because of all the moving, I think. Like, I was over at this girl's house and she just moved in and we had them over for dinner once because her husband's a missileer. We'd talk at church and that was kind of the extent. And then our husbands just happened to both be out on alert and she was like, "Oh, you should come over. I'll show you the house." You know, I was like, "Yeah. Great!" And so we went over and we were talking and I realized, like, I was asking her questions that, I mean

they're not super, super personal but I was like, "Oh, yeah, so what about this, so what about this?" And we're talking about things as if we had been friends for a long time . . . and I thought, we don't have six months to talk about "What's your favorite movie?" and things like that because we only are going to live together for four years at the most. And we've just got to do it quick because you never know when you are going to need a ride to the airport at 6:30 a.m. like she did. She knew that I get up early, so it was easy for her to ask me, and I didn't mind.

There's a sense of neighborliness in Lindsay's comments—a sense of people knowing they can rely on each other—that may seem more suited to a lifetime of close communication than the ostensibly hit-and-run connections of military life. That is an interesting irony, and an undeniable one. For military children and their mothers alike, the unpredictable transiency of post-to-post life seems to intensify, rather than diminish, personal relationships.

"THE WORST PART OF IT": DEPLOYMENT

For most military wives, the greatest stress—far greater than that caused by constant moving—comes in those periods when they are safe at home with their families while their husbands are deployed, sometimes in harm's way. My assistant, Kimberly Keck, and I asked every woman we interviewed what was the "worst part" of being a military wife; I can't remember one who didn't say respond with some version of Jean Harrison's comment, "When your husband goes away." Virtually all of them would agree that, whatever difficulties might exist on the home front, having your spouse home and "not getting shot at" is, in Harrison's words, "a great blessing to have."

The anxiety associated with the enforced absences of deployments came through in a telling slip of the tongue when Kimberly interviewed Cara Amsden, whose husband, Nathaniel, was at the time an Air Force ROTC cadet. In explaining her initial reluctance to marry into the military, Cara admitted she would find it hard to take "those times when they're, you know—oh, I just—" She searched unsuccessfully for the right word until Kim, herself the daughter of a long-term member of the Army National Guard, hesitantly offered, "Deported?" Both of them knew that wasn't right, and by the end of the interview they remembered the correct term: "deployed." But the temporary mix-up spoke volumes about how wives of warriors feel when their husbands are sent by the order of a faceless government to some far-off land, perhaps never to return.

During deployments, wives miss having their husbands around on a daily basis, and they also worry constantly about their safety. When asked about the time that her husband, Mike, was deployed in Afghanistan, Jessica Richins replied that she had "a whole lot of anxiety attacks. When he left, we were not engaged. We were dating, wanted to get married, but I told him he was not allowed to propose because I couldn't deal with being engaged for a year with him in Afghanistan. It was just too stressful." While not all military sweethearts would postpone an engagement as Richins did, the emotional turmoil she describes is common to many.

For wives with children, there is a special anxiety related not just to missing their husbands, but also to what their husbands are missing in being apart from the family. "I have a tremendous respect for those who serve in the military," says Tamara Johnson, "because they do make tremendous sacrifices—they miss birthdays, they miss anniversaries, they miss births, they miss, you know, those kind of things in their lives."

Keri Hatch is eloquent in describing the corrosive effects of enforced absences. Even as she expresses pride in her husband's service, she laments the emotional cost that extended deployments placed on her family. She is speaking about multiple separations from her husband, Rich, during deployments in the 1980s.

> I guess that would probably be the highest high and the lowest low in a way happening at the same time, because it was such a noble cause . . . so perhaps that's a little bit of history, where people go in giving their very best and then also end up sacrificing family time and the things that that person who's deployed misses in terms of your family, in terms of graduations and just life. When Rich was gone the second time, our son won the state championship in his relay . . . and he missed that and he missed my two oldest kids graduate from BYU . . . He missed all of Jared's senior year, all of Tanner's freshman year, the birth of our grandson . . . The deployment is a pretty low time.

Even when the military makes a special effort to unite wives and husbands for extraordinary events, the reunion is generally as short as it is sweet—and that can create a special kind of frustration. Sarah Jones, for example, is married to an Army National Guard second lieutenant, Chad Jones. He was deployed to Iraq during her first pregnancy and nearly missed their oldest child's birth. She describes the ups and downs of that difficult period:

When we found out he was deploying that was a total surprise to us because he deployed with a different unit than he's actually assigned to . . . So that was a total shock, it was a little overwhelming at first. Once he did leave, I wasn't too bad with him leaving the first time . . . And then he got to come home when Josh was born and we . . . had to stay one full day at the hospital . . . So we were able to spend one night at home together as a family, and that was really good . . .

I don't think it hit either of us, really, how long he was going to be gone, until we got to the airport for him to leave after Josh was born. That was really hard. Because that's when it finally hit me that, "Oh, my gosh, I've got a newborn baby that I have to take care of for the next year while he's gone." . . . And after he left after Josh was born—I think most Army wives do this—you just kind of go into survival mode and sometimes you just do what you have to do to make everything work, make like the best you can while your spouse is deployed. And that's pretty much what I did . . . The reality of it I don't think really hit me until Chad got home and I looked back on it. Because the whole time he was gone I was just doing the best I could, just putting one foot in front of the other . . . I didn't realize the full extent of it just until after he got back home and I said, "Holy cow, that was rough."

Even wives who face the challenges head on, with a brave official face, can have moments of separation anxiety. Kaylee Folsom shares the following story:

I'm a singer, and my major is music education, and one of the pieces I was learning is from an opera by Puccini called *Madame Butterfly*. And it's about this geisha who marries an American sailor and he leaves, and says he's going to come back, but he doesn't come and he doesn't come and he doesn't come. And her servant and everybody else keeps telling her, "Give it up, he's not coming back." And she sings this piece about her dream of what it's going to be like when he comes again. And that his white sails on the horizon, and that should mean that he's coming up the hill and calling her by name the way he did before. And, at the end of the piece, she says, "You must believe that this will be so, I stand with faith secure, awaiting."

And the power of the piece itself really hit home for me. I realized this piece, this character, is me. This is me, and I, I don't know, I just realized how much I was aching to feel my husband's arms around me and so I, in the middle of my voice lesson, I just started bawling and my poor voice teacher, she was really patient, but I've never cried so hard in my life; I huddled in a corner and hugged my knees to my chest, but after that cry I went home and I said a prayer thanking God for all of the blessings that he's given me and for the opportunity to be at school finishing my education and to be a part of this lifestyle that I am a part of. Because it's changed me, it's made me a better person and I think that's irreplaceable.

Kaylee's emotional recollection summarizes vividly the stressful effects of separation during deployments. But it also reveals the ability of wives left at home to confront the stress, to hope for the best, and to lean on a faith in God that is central to their worldview. I'll speak more in a moment about this important aspect of military wives' coping strategy.

DISCONNECTION AT HOME

When husbands return from deployment, there is naturally relief and happiness. But often there is a distance that remains between husband and wife—a distance that can seem to extend the absence even after it is physically closed. Some husbands have a hard time leaving the war behind and interacting well with their families, and for others it is a matter of being unable to speak about matters that the government has classified. When John Madsen was being trained as a missileer, his study materials were classified and kept in a locked briefcase: he was forbidden to even look at them if Lindsay was in the same room. Now that he goes on twenty-four-hour alerts, there are still classified elements that he cannot talk about. Even if something about missiles is leaked on the news, he is unable to answer his wife's questions. Spouses in circumstances like Lindsay's—and there are many of them—therefore have to contend with whole blocked-out areas, and this can be troubling to a relationship that depends on trustful communication. Melissa Keck, acknowledging ruefully that "there are a lot of wives who don't know where their husband is, don't know what he does for a living," admits that such official barriers can be problematic; but, with a shrug, she says, "You get used to it."

A more serious type of disengagement occurs in cases where a spouse returns from a war zone unwilling to "debrief" his wife on what he has been through—or even to talk about experiences that may have been traumatic. Such a response is of course not uncommon among soldiers returning from combat, and a bulky literature exists on such disengaged behavior, both in its milder forms and in the full-blown version called post-traumatic stress disorder.

Keri Hatch's husband was deployed twice to Iraq within a relatively short period of time. When he came back, he felt a need to always be able to physically escape an area. "He felt very vulnerable, and he felt like he couldn't protect us. And I would catch him—when we'd go into a building, he would like check it out to see how we could escape, or how he could protect us. I mean that's the kind of environment he had come from. It was like, okay, when you walk into a building or you walk into an area, where can I go to get safety? And so, he came—that's just like a very small example, but, oh, he couldn't even do fireworks." Her husband had not been in direct combat. As a member of JAG, he was there helping the Iraqis write a new constitution. Hatch continues: "As far as our marriage, it was a very difficult time. There was probably one period of time when I just, you know, he's been home for—it was like the end of the summer and I finally just said to him, 'You know, do you want the boys and me to leave?' You know, like not, you know, not divorce, but just, like give his some space and let him be on his own. And he and I think that was kind of his wake-up call, that he needed to be more in touch with us."

Keri and Rich, with the help of their church, were able to work through this problem and restore their connection. In recalling that difficult period, however, Keri still became emotional and had to take a moment to collect herself. Her reaction, it's worth mentioning, is a permissible reaction for a woman speaking to another woman—and this points to another element of military wives' worldview that, in my estimation, gives them a type of strength that is denied their husbands. Their willingness to be emotionally vulnerable gives them a kind of flexibility and openness to complexity that, just like resourcefulness, helps them to cope with difficult situations.

"A TIGHT-KNIT COMMUNITY"

If the women we interviewed see deployment as the worst part of being married to the military, almost all of them say that the best part is the friendship and support they get from other military wives. That connection, they understand, comes from sharing experiences that are so nearly universal among them that they don't have to be explained. Keri Hatch puts it well when she says that, in conversations with other military spouses, "you don't have to go into a lot of background information, you don't have to explain where you're coming from. They just get it." The "it" here includes the whole gamut of exhilarating and trying emotions that comes with being the spouse of a service member. It is that "it," as much as anything, that informs these women's attitudes toward friendship and that makes them see themselves as members of a community.

I say "community" because that word comes up frequently when these women describe their lives as military spouses—and indeed, when they describe the military itself. "The military," says Keri Hatch, "is a very tight-knit community, and people take care of each other." That is a typical observation for these women, as is Alicia LaDuke's comment about her experience living on base: "I like just being near, knowing that my neighbors are all military and they all know what's going on with each other and you just have this sense of everyone looking out for each other and sense of community." Sarah Jones expresses a similar viewpoint when she says, "Just knowing that there's somebody else who can relate to you and who you could turn to if you ever needed to is a huge support."

In all of these comments, community is understood as a place where people watch out for each other. That "watching out" often includes such practices as sharing child rearing, hosting community meals and play parties for the children, and inviting newcomers to a base over for dinner—something that my daughter Lindsay and many others consider almost a moral obligation. Keri Hatch has served as a coach for women delivering babies, and as a temporary foster parent for teenagers finishing high school when their parents were abruptly transferred. Such practices are seen as almost the normal requirements of being a good neighbor. But there's a protective element to "watching out," too, that is perhaps analogous to the protectiveness that warriors in combat are often said to feel for each other. So caring for each other is a consciously acknowledged folk unit of these women's worldview.

In describing its importance, several women use the term *camaraderie*. Jean Harrison, for example, notes that "the things that you're going through, your neighbors are going through. That's great because that builds up that camaraderie." Lori Sturgeon, reflecting on the differences between living on base and "on the economy" (that is, in the civilian population), notes that off base "you usually will get a bigger house for your money, but you don't have the camaraderie . . . In the military base, you never felt like you had to go out of your way, knocking on doors, to make friends; you just instantly had friends." Like *community*, the word *camaraderie* is an interesting and appropriate designation. It suggests, literally, the closeness of comrades—that is, of fellow warriors in a fight of "us" against "them." So there may be an element of quiet pride, and even of defensiveness, in these women's view of themselves as military spouses: to the extent that they are comrades to each other, they are like their warrior husbands—and unlike civilians. And like good comrades, they support each other in their mutually experienced dislocations. They become almost each other's surrogate families, standing in for their loved ones far away.

When we asked military wives what the best thing about the military was, we got a variety of answers. Some liked the inexpensive medical care. Others chose the lessons, clubs, or inexpensive commissary shopping available on base. Some even favored the military balls. But the most common answer was the people, the friends that they were able to make as part of the military community. The spouses find in each other the support and friendship that makes military life worthwhile. If the military tends to "dislocate" spouses, then it is their contact with each other that "relocates" them. A critical component of these spouses' worldview is the ability and expectation of finding friends wherever they are.

"AMAZING PEOPLE"

If military wives gain important emotional support from their personal interactions with other wives, they gain a more public type of support from their feelings about duty and patriotism. Love of country, coupled with an absolute respect for their husbands as defenders of the country, comprises an important second element in their common worldview, and one that leads them, even in the most difficult circumstances, to feel pride as well as comfort.

Patriotism, of course, is a very public matter at military installations, where flags fly proudly and the national anthem and "Taps" are played each day over base-wide loudspeakers. At Malmstrom Air Force Base, those in uniform are required to stand at attention as each selection is played, and when "The Star-Spangled Banner" is played, people outside stop what they are doing, place their hands over their hearts, and listen respectfully. So pervasive is the sense of respect accorded this most patriotic of American songs that even children learn to adopt it at an early age. When my grandson Caleb was only two and a half, he was already placing his hand over his heart during what he called the "outside song." In such an atmosphere—and given the fact that military bases are staging areas for deployments—it is hardly surprising that an avowed love of country is a principal ingredient in military spouses' worldviews.

This is not to say that every military wife—or for that matter every active-duty service member—agrees with every aspect of US government policy. As in any other community—or at least any other democratic community—opinions about the wisdom of this or that order from above can vary from docile acceptance to irate objection: contrary to the antimilitary conventional wisdom, members of the armed services are no more robotically compliant than the members of any other bureaucratic institution—and this goes for military wives as well as their husbands. Few wives, however, are comfortable criticizing missions that their husbands are currently involved in; doing so might be seen, legitimately or not, as amplifying the danger that the troops are already facing. So, like their husbands, military spouses can have a hard time when outsiders—those "on the economy"—criticize what their husbands are doing; their patriotism, to a certain extent, is inseparable from marital loyalty. So, whatever their views of a given mission, they tend to support the marching orders that their men have to follow. Some of Jessica Richins's friends, for example, are opposed to the current war. "Which is fine," she says. "But I don't feel like you have to support war to support the troops." And supporting the troops—their loved ones—is the heart of their patriotism.

But that support doesn't emerge in a value-free vacuum. When these women express patriotic pride in what their husbands are doing, they often speak specifically about the sense of sacrifice that devotion to a military career can entail. That sense can encompass something as relatively mild as missing a child's birthday, or as grave as what we tritely call the "ultimate sacrifice." Whatever the specifics, the wives who spoke to us comment frequently on how impressed they are with their husbands' willingness to undergo trials and put themselves at risk for the sake of others. Kaylee Folsom's husband, Aaron, for example, graduated from West Point. When asked about his decision to attend the military academy instead of a civilian university, she replies: "He decided to go to West Point because he wanted more structure. He wanted a challenge, and he is very patriotic, he loves his country, and he wanted to have the experience of being able to be a part of something that's bigger than yourself. He honestly wanted to be able to make the sacrifices that he's making so that other people wouldn't have to. And he decided to do the infantry before medical school so that he could relate to the soldiers that he'll be treating someday."

Kaylee's respect for what her husband is doing is profound, and in her choice of words she reveals that for her, as for many other wives, what is most worthy of respect is service to others. Just as they value caring for each other when their husbands are away, military wives deeply respect their husbands' caring for the nation at large. As a fundamental value, "watching out" for others works on both the personal and the civic level.

They see that value in their husbands, to be sure, but also in the military world at large, and to some of them its appearance as an institutional value is more than merely impressive; it's "amazing." "I think the military experience is amazing," says Tamara Johnson, "for those who are doing it for the right reasons: that they want to serve their country." Keri Hatch, referring to her husband's comrades in the Middle East, calls them "just an amazing group of people." Jessica Richins echoes that sentiment more effusively. Acknowledging again that "people" are the best thing about the military experience, she speaks about her husband Mike's fellow soldiers: "I love the group of guys he went to Afghanistan with. They are stand-up guys, they all are. They're the most amazing men, and their families are amazing too . . . The people that are in the military are awesome. They're really quality and that they're willing to lay down their lives for our country is amazing. And so, to be able to be around so many of those types of people is awesome. I love it."

During our interviews, several of the women were moved to tears or mentioned times when they had been so moved as they discussed their husband's sacrifices for their country. Unlike their husbands, who are trained to be stoic, they are willing to demonstrate publicly their emotional attachment to patriotism and its attendant sacrifices. This willingness is evident in Kaylee Folsom's touching anecdote about her *Madame Butterfly* epiphany and in Jean Harrison's memory of tearing up during an Army Reserves recitation of the Soldier's Creed. It comes through, too, in Lori Sturgeon's recollection of her own emotional evolution from clueless bride to mature military spouse: "When I first got married, I would see all these older women . . . when that flag would come out, their tears would come down their faces and I would be like, "Why are you crying? I don't understand." But now I totally get it. You know, the sacrifice, the dedication to the people who have died in service. What that flag comes up, I can't, even now, help but cry, because I think of all the people who have lost their spouses and sons, and it just really hits hard." To Lori—the wife of a retired Army lieutenant colonel—such expressions of emotion, far from being signs of weakness, are certifications of the depth of feeling that she and other wives have for the service. They are the visible signal of an invisible affection that is at the very heart of these women's worldview.

"GOD AND HIS COUNTRY"

The final piece of these women's coping worldview is their religious faith—specifically, their faith as members of the Church of Jesus Christ of Latter-day Saints. It would perhaps be more accurate to call this the first rather than the final piece of their worldview, because to them religion informs everything they do, and it nurtures both their patriotism and their camaraderie. These women are not lukewarm or part-time believers, and the God they worship is not—to quote an old song—"the kind you have to wind up on Sunday." When we asked them what role religion played in their lives, the answers were remarkably consistent. "Religion is my rock," says Keri Hatch. In hard times, she is consoled by her conviction that "the Army's not in charge; Heavenly Father is in charge." To Kaylee Folsom, love of God is "the focus of everything I do." To Melissa Keck, religion is "how I would mainly define myself. Before I'm an Army wife or a volunteer at the school, or anything, I'm a member of the LDS Church . . . It kind of grounds me and that's who I am."

Because religion is so central to these women's identity, it functions as an invaluable resource in coping with difficulties. In fact, many of the wives we interviewed admitted that, without their religion, the military lifestyle might not have been either attractive or sustainable. "Without my religious beliefs and my testimony," says Keri Hatch, "I don't know that I could have survived a military life." Sarah Taylor-Jenson, reflecting on her struggles as a single parent when her husband was deployed, says, "I wouldn't have been able to do it without my religion." And Kaylee Folsom, acknowledging that her faith is "everything to me," calls it "the reason why I'm happy even though my husband isn't with me. I'm happy because I know that God is watching over him and he's watching over me." In this comment we hear again the respect paid to caring—and we see here that it stems from a conviction that the implicit model for the perfect caretaker is God himself.

One reason that these women find the military a congenial environment is that they perceive it as welcoming and nurturing their religious sensibilities. They see the military as similar in spirit to a faith community, a place where belief in a higher power is common, where service to others is an honored value, and where people are more serious about their beliefs than the civilian population. "We have met deeply religious people in the military," says Keri Hatch, "who are very devoted to their beliefs and to a Christian lifestyle . . . In general, military people, if they are religious, are very true to their beliefs." Kaylee Folsom agrees, and suggests that the military, beyond merely welcoming people of faith, may actually create faith in those who didn't possess it before. In an interesting spin on the old adage about there being no atheists in foxholes, she says, "The people that I know that aren't necessarily religious, after they have experiences in the Army, they become more religious, or at least want to believe that there is a higher being looking out for them . . . The people that I know that are religious, whether they're Latter-day Saints or not, are absolutely firm in their faith, and they live by that more than anything. I've met some of the most spiritual people in the military."

Whatever the validity of their perception may be, it's clear that, to these wives, it is no accident that military communities tend to be more religious than civilian ones. Cara Amsden, noting that the Air Force seems to have a high concentration of LDS families, suggests that Mormons may be attracted to the service "because we feel more of a responsibility for what we've been given from the Lord. To protect our freedoms and our family and the things that we've received from the Lord, we feel more of a responsibility to fight for those things." Lori Sturgeon make a similar point. In speaking of her husband, she says proudly that he does his job "for God and his country, and what more valiant thing can you do than that?" "For God and Country" is of course a time-honored motto of the military services—one that may be particularly appropriate in a traditionally Christian country that one historian has dubbed a "redeemer nation" (Tuveson 1968). To women like Cara and Lori, donning the uniform of such a nation is an act deserving of both pride and devotion. As I've mentioned earlier, they may not agree with every government policy, but belief in the basic righteousness of the American armed forces is a driving element of their spiritual—and political—worldview. Holding on to that is another way in which they are able to cope.

FOR FUTURE EXPLORATION

Military wives face challenges that, to married people not in the military, might seem daunting, even terrifying. I don't deny—nor do the wives we've spoken to deny—that the challenges are severe. Even when their spouses are stationed "at home," they must contend with long hours, unpredictable schedules, frequent relocations, and a traditional formality that, to most young American women, hardly comes naturally. When their mates are deployed, these problems are aggravated by enforced separation and the anxiety of not knowing when, or if, they will ever return. I have tried to show in this chapter that, despite these burdens, military wives tend to look favorably on their experience, and that they employ as a quite effective coping mechanism a worldview that honors friendship, patriotism, and faith.

There are probably more questions raised than answers supplied by my exploration of these spouses' worldview, and I offer it not as a definitive study but as a spur to further investigation. It would be interesting to know, for example, whether the values expressed by these women—especially the pro-military values—were acquired mostly in childhood and adolescence or whether, like Lori Sturgeon's "learning to cry" at flag raisings, they came to fruition only after they married into the military. In other words, are young women "self-selected" to be military wives, or do they grow into the role once they are "selected" by their husbands?

We might also ask whether the process of making lemonade from lemons—that is, the women's creative adaptation to difficult circumstances—is simply a matter of making a virtue of necessity or whether there is some inherent psychological "lift" that comes from wresting pleasure, or at least satisfaction, from adversity. Critiques like those of Weinstein and White tend to suggest that wives' "coping" mechanisms are mostly an unwelcome burden, seen as unavoidable by "dutiful" spouses. That may be true. But we need more data—that is, military women's own words—to determine whether "joyful labor" is not also part of the picture.

It would also be useful to investigate whether the coping worldview of these LDS women is shared by differently focused military spouses. Do we see the same cluster of camaraderie, patriotism, and spirituality among military wives belonging to other Christian denominations? Among wives whose allegiance is to Jewish or Muslim or other non-Christian religious traditions? What about that small but growing cadre of military spouses, sweethearts, and partners who are men, not women? Or the recently emergent community of gay and lesbian partners? Do the LDS accommodations described here function with the same efficacy among these other groups? Or do those groups cope in different ways?

Finally, we might examine more closely the relationship between military spouses' worldview and that of the personnel whose service they "support." In reviewing the transcripts that Kimberly Keck and I collected for this project, I have been struck by how frequently a wife's expression of a cherished "folk idea" suggested parallels to certain warrior worldviews. One major difference between wives and warriors in this regard, of course, is that the former are allowed to cry and the latter are not. But in other respects the parallels are often close.

"Camaraderie," for example, which so many of these women applaud, is essentially the military notion of warrior comradeship. Patriotism—especially its injunction that one must not undermine the troops by damning their mission—could be as easily endorsed by a general as by a military wife. And religion? Religion has long been used as a warriors' rallying cry, not always in an admirable context. It would be useful to know to what degree Lori Sturgeon's "God and country" equation would be endorsed by her husband or the men he commands. (It is, not so incidentally, endorsed by the US Army as the official motto of the Chaplain Corps.)

It is often noted that military culture—despite the entry of women into the armed services—remains largely a masculine preserve, awash in the testosterone-driven aggression that "masculine" implies. The values of the women examined here, by contrast, are anything but aggressive; they are, in the best tradition of "gentlemanly" or "ladylike" behavior, those of common decency and respect for others. One may agree or disagree that military institutions in general honor those values. But that they animate these women's lives seems beyond dispute. And to the extent that these "shadow enlistees" are in reality if not in name part of the military, it may make sense to consider their worldview not merely a "female" coping mechanism, but a cluster of "soft" cultural values that is as important as masculine toughness in defining a warrior ideal for American society.

WORKS CITED

Amsden, Cara. 2011. Personal interview with Kimberly Keck.

Dundes, Alan. 1971. "Folk Ideas as Units of Worldview." *Journal of American Folklore* 84 (331): 93–103. http://dx.doi.org/10.2307/539737.

Folsom, Kaylee. 2010. Personal interview with Kristi Young, September 24.

Harrison, Deborah, and Lucie Laliberté. 1997. "Gender, the Military, and Military Family Support." In *Wives and Warriors: Women and the Military in the United States and Canada,* ed. Laurie Weinstein and Christie C. White, 35–54. Westport, CT: Bergin & Garvey.

Harrison, Jean. 2010. Personal interview with Kimberly Keck, November 27.

Hatch, Keri. 2011. Personal interview with Kimberly Keck, January 19.

Huebner, Angela J., Jay A. Mancini, Gary L. Bowen, and Dennis K. Orthner. 2009. "Shadowed by War: Building Community Capacity to Support Military Families." *Family Relations* 58 (2): 216–28. http://dx.doi.org/10.1111/j.1741-3729.2008.00548.x.

Johnson, Tamara. 2010. Personal interview with Kristi Young, October 19.

Jones, Sarah. 2011. Personal interview with Kristi Young, July 28.

Keck, Melissa. 2010. Personal interview with Kimberly Keck, December 30.

LaDuke, Alicia. 2010. Personal interview with Kristi Young, September 15.

Madsen, Lindsay. 2010. Personal interview with Kristi Young, September 14.

Richins, Jessica. 2010. Personal interview with Kimberly Keck, September 24.

Slik, Debbie. 2010. Personal interview with Kristi Young, November 9.

Sturgeon, Lori. N.d. Personal interview with Kimberly Keck.

Taylor-Jenson, Sarah. 2010. Personal interview with Kristi Young, September 24.

Tuveson, Ernest Lee. 1968. *Redeemer Nation: The Idea of America's Millennial Role*. Chicago: University of Chicago Press.

Weinstein, Laurie, and Christie C. White, eds. 1997. *Wives and Warriors: Women and the Military in the United States and Canada*. Westport, CT: Bergin & Garvey.

9

Oppositional Positioning
The Military Identification of Young Antiwar Veterans

Lisa Gilman

On November 4, 2008, the same day that Barack Obama was elected president of the United States, a group of young American veterans, most of whom had served in the current conflict in Iraq, opened Coffee Strong, an antiwar coffeehouse and nonprofit organization located outside of two military bases in Lakewood, Washington, which have since been joined into Joint Base Lewis-McChord (JBLM). Though the veterans identify themselves in opposition to the US military and its current conflicts, the culture of their activist community is rife with military folklore and folklore that plays ironically with military symbols to create multiple messages. The name Coffee Strong, for example, is a play on the US Army slogan, "Army Strong." Coffee Strong members wear fragments of their military uniforms, display symbols that reference the military, and tell jokes, legends, and personal-experience narratives about the military, war, and Veterans Affairs. Their stories and conversations are replete with esoteric knowledge, acronyms, and the type of folk speech that Elinor Levy, in chapter 5 of this volume, terms "enlistic." This essay explores why this antiwar community relies so heavily on American military folklore in its group formation, identification, and activism.

In his influential article "Differential Identity and the Social Base of Folklore," Richard Bauman (2000 [1972]) pointed out that folklore performance is not always an intragroup phenomenon, but may sometimes be based on differences in identity. More recently, Dorothy Noyes (2003) has convincingly demonstrated how folk groups are interactively constituted.

DOI: 10.7330/9780874219043.c09   181

In this essay, building on these conceptualizations of group and identity, I examine how a folk group's identity can be complicatedly founded in the simultaneous identification with and rejection of (or differentiation from) a larger cultural community of which they are also members. Through their performances of folklore associated with the military, the activist war veterans of Coffee Strong identify themselves as members of the larger military culture at the same time that they politically reject this association; their performances of military-themed folklore are expressions of both identification and differentiation. These oppositionally positioned and sometimes ambiguous performances also function interactively with people outside the activist group to communicate multivalent messages, advance the group's objectives, draw people into the movement, and further differentiate the members of Coffee Strong from those critical of them.

COFFEE STRONG, COMMUNITY, AND THE
COMPLEXITY OF IDENTITY

This research began as a documentary film project, *Grounds for Resistance,* about activist veterans involved in Coffee Strong. I completed the filming for that project in March 2010, but my fieldwork has continued through the postproduction phase and into the present. In my research for the film, and in compiling the material used for this essay, I took an ethnographic approach, establishing friendships with many of those featured, spending time in the coffee shop participating and observing, and attending as many of its sponsored events as I could. Much of the data presented here comes from my participant observation, which was largely unrecorded and thus is presented as descriptions rather than verbatim transcripts. Those parts that were documented in film are presented verbatim when relevant and possible. Given the personal nature of some of the material in this essay, I refer to veterans using only their first names. For those who indicated that they did not want to be identified, I include only limited information about their ages and the branches in which they served.

Coffee Strong is located in Lakewood, Washington, between Olympia and Tacoma along the Interstate 5 corridor within sight of the security fences that enclose Fort Lewis Army Base and Lewis-McChord Air Force Base (now JBLM). It shares a wall and parking lot with its neighbor, a Subway sandwich shop, one of many fast-food restaurants on this one-mile strip, which resembles those in many towns that cater to military bases (figure 9.1).

9.1. The Coffee Strong storefront.

The predominant businesses are fast-food chain restaurants, pawn-shops, barbershops advertising military-appropriate haircuts, and military surplus stores. Coffee Strong stands out in this military enclave. Locally owned by veterans, it offers the expected menu and atmosphere of contemporary coffee shops. It features high-quality coffee—served straight and in lattes, cappuccinos, and mochas—in addition to other food and beverage offerings. It provides computers, free wireless Internet service, chairs and tables, a nook with a cozy couch and cushioned chairs, and a foosball table. In contrast to other businesses in the neighborhood, its walls are plastered with political artwork and posters, and above the coffee counter is a display of hand-printed antiwar T-shirts, designed by Coffee Strong members, that are available for purchase. A bookshelf holds radical leftist and antiwar literature that customers are free to read, borrow for an indefinite time, or swap with their own books. A table hosts brochures that advertise resources available to active-duty and veteran military personnel—flyers about counseling opportunities, how to navigate the complex world of veteran benefits, statistics on military sexual trauma, and how to file for conscientious objector status.

Coffee Strong founders were inspired by the coffeehouses, located outside numerous military bases, that were important to the GI movement during the Vietnam War. These coffeehouses, staffed by civilian activists and active-duty personnel, provided places away from bases where military folks could assemble to talk about the war and associated political issues, obtain legal counseling, and organize political actions. Some also published provocative newspapers that they infiltrated onto bases (Cortright 1975; Moser 1996; Nicosia 2001; Norris 2008). The founders of Coffee Strong, young American war veterans who served in Operation Iraqi Freedom (2003–present), emerged from their war experiences politically, socially, and emotionally transformed. Many of them met while pursuing undergraduate degrees at Evergreen State College, located in Olympia, Washington, twenty miles from the base. Having learned about the Vietnam War–era coffeehouse movement largely from watching David Zeiger's documentary *Sir, No Sir!* and battling with varying levels of combat stress, guilt about their participation in a war that they had come to feel was unjust, and memories of what they had seen and done, these young veterans created a nonprofit that allows them to integrate their emotional responses and their transformed political identities into a project that provides important services and resources to military personnel and veterans, creates a community for veterans to deal with their experiences, and serves as a center from which to participate in regional and national antiwar activism.

Coffee Strong, a project of the nonprofit GI Voice, is a coffeehouse that sells coffee (actually, it gives it away to anyone with military identification) and also provides a place for active-duty military personnel, veterans, military family members, and civilians to read, discuss their military experiences, and hold political meetings. The local chapter of Iraq Veterans against the War, to which most of the Coffee Strong veterans belong, uses the space for its meetings after hours, and the space has also been used to train GI rights counselors and advocates for victims of military sexual trauma. The Coffee Strong community also uses the space for nonpolitical- and nonbusiness-related activities, such as birthday celebrations and other parties. These events, though not overtly part of the mission of the nonprofit or business side of Coffee Strong, nevertheless contribute to the forging of interpersonal bonds and raise awareness about the organization's activities by bringing new people into the space.

Coffee Strong, then, is a space where individuals hang out and where, in their interactions and shared activities, they form a community. As activist Joseph explains, "Coffee Strong isn't the community, it's just a space around which we organize a community" (for discussions of similar concepts, see

Bauman 2000 [1972]; Fine 1979; Noyes 2003). The Coffee Strong community, heterogeneous and fluid, is constituted through a combination of coming together in the space of the coffee shop and participating in its many activities. No formal process exists for designating who is and who is not a member, and to Coffee Strong participants the question of "official" membership is considered largely irrelevant because inclusivity and porous group boundaries are central to the organization's self-conception.

Included in my definition of the Coffee Strong group for purposes of this essay are members of its board (veteran and civilian activists), the people who volunteer and work at Coffee Strong (veterans and civilians), and the active-duty soldiers, veterans, and civilians who spend time in Coffee Strong, participate in its activities, attend meetings, and take advantage of its services. Not included are those customers, some of them regulars, who come to Coffee Strong simply to drink coffee, interacting minimally with the activist participants. My focus for this essay is on the young war veteran members, most of whom are no longer in the military, as they have been most influential in shaping elements of the group's culture, such as organizational focus, décor, activities, oral tradition, and selection and design of visual symbols. I refer to them as "veteran activists" to distinguish them from the civilian participants. The veterans are also the ones most relevant to my analysis of the simultaneous identification with and differentiation from the military.

Following Dorothy Noyes (2003), the Coffee Strong group can be seen as a "multiplex network," bringing together heterogeneous individuals who are differentially networked to others within the group as well as to external networks, which in this case includes active-duty military personnel. In contrast to Coffee Strong, membership in the military is exclusive, rigidly bound, and often eternal. People officially join the military and go through a rigorous training process through which they learn the cultural codes of the community (see Burke 1989). After they are discharged, many continue to identify with the military on a symbolic level in addition to benefitting tangibly from their access to Veterans Affairs and other military resources. Like other veterans, Coffee Strong veterans are members of the larger military network in an official capacity, and through their past—and, for many, continued—participation with Veterans Affairs and others involved in this network, they have access to esoteric knowledge, beliefs, folk speech, legends, anecdotes, and other forms of military folklore. It is this complexity of identification with multiple networks that creates the possibility for the simultaneous identification and differentiation that define the activism of the veterans involved in Coffee Strong.

The Coffee Strong veterans forge bonds among themselves based on this strong association with an organization that they now reject. In this they are similar to other groups whose membership criterion is based on a denunciation of a previous experience, identity, or behavior. For example, Carole Cain explains that qualification for membership in Alcoholics Anonymous is "a negation of the behavior that first qualified one for membership" (1991, 6). The main thing that members in these types of groups have in common is that they shared a behavior that they now reject; it is their very connection with what they renounce that is the central identifier for the group.

For the veteran activists, their military experience is at the core of many aspects of their current lives. In their college courses and independent reading and research, many study war, imperialism, and activism. They spend much of their time at Coffee Strong hanging out with other veterans and engaging in antiwar activism. Many of their friends are veterans, and most are planning their futures in relationship to their war experiences. Some plan to continue to be activists, others are training to be teachers and hope to educate their students about the realities of war and imperialism, and others hope to work for social service agencies that serve veterans. Their identification with the military is therefore integral to their personal sense of identity, even as in many ways they position themselves in opposition to the institution. I should clarify, however, that each individual involved in Coffee Strong has his or her own unique, complicated, and evolving way of thinking about his or her relationship to the military and war. In this essay, I try to analyze the complicated process of oppositional positioning without essentializing any given individual's perspective or feelings at any given time.

Within the wider Coffee Strong community—that is, the community that includes active-duty military personnel—it is the young veterans' firsthand military experience that qualifies them to speak authoritatively on the problems of war and militarism; their insider status in many ways makes them better positioned than civilian activists to engage with active-duty troops on their own terms. As Joseph explained in an e-mail to me on August 31, 2011, "Many of us are aware that our status as veterans imparts a certain privilege to speak, and that we very deliberately invoke this privilege in order to reach an audience that may not otherwise want to hear what we have to say about war, PTSD [post-traumatic stress disorder], and military service." Not all the veterans involved, however, feel comfortable with this positioning. One activist, for example, explained to me in a conversation on September 2, 2011, that during his time working full-time at Coffee Strong, he had wanted to distance himself from his veteran status: he wanted to be identified as himself rather than as someone associated with the military. Nevertheless, he felt compelled to publicly present his association with the military because in this context, he felt that his experience in war justified his activism.

The veteran activists have strategically created a physical space that visually expresses this oppositional relationship to the military. The paradox is evident even from the street. On the windows are the painted words "veteran-owned" and a sign advertising free coffee for those with military identification, both of which are indicators of the business's association with the military. Yet next to these notices are flyers advertising hard-core punk shows, screenings of radical films, and resources for survivors of sexual trauma. This combination of signage communicates messages that are ambiguous enough that someone not familiar with the organization might interpret them as pro-military and enter Coffee Strong oblivious to its political positioning. One active-duty soldier I met at Coffee Strong, for example, explained that he did not drop by Coffee Strong for a long time because the sign and name led him to assume that it was a hangout for officers, not his preferred company during his off-base time. By contrast, those familiar with the organization understand that encoded in all the flyers is Coffee Strong's criticism of the military. For example, the sexual trauma flyer references Coffee Strong members' perception that military sexual trauma is a rampant but largely ignored problem, and that as a result nonmilitary organizations such as theirs are needed to provide services to the too many victims.

Coffee Strong's physical positioning so close to the base and establishments popular with soldiers is in itself an announcement of its paradoxical relationship to the military. The founders of Coffee Strong strategically selected this location because it defied what they perceive to be the military's attempts to exert control over its members' freedom of expression and political activities, and because proximity to the base allows them to more easily provide resources to military personnel. Coffee Strong's paradoxical positioning is highlighted during mealtimes, when streams of soldiers in their desert camouflage uniforms pass in front of Coffee Strong and walk into the adjacent Subway sandwich shop. A few drop by to pick up coffee or even to stay to drink it, though many glance at the posters on the windows of Coffee Strong and pass by, some making comments about what people would think if they entered. The comments of those soldiers who scoff and walk on suggest that they recognize the irony of Coffee Strong's presence in a locale saturated with military culture.

When I first visited Coffee Strong, I was struck by the veterans' choice to spend so much time near a military base when they were struggling with issues related to their war experience. I assumed that in a similar position, I probably would choose to be as far away from a military base as possible, and certainly many veterans do make this choice. What I had not recognized yet was how central their military experience was to their current sense of themselves, or how deeply—despite their issues with the military—many continued to identify with other soldiers. Joseph explains the conflict especially well: "I don't like certain features of military culture—like heterosexism, gender discrimination, racism (especially towards people of assumed Middle Eastern descent) and hyperaggressive behavior—but I can easily identify with the common soldier because we share similar experiences." Based on my observations and conversations, I also think that being within a military milieu while having the freedom to outwardly defy its authority has been for many of these veterans an important part of their healing processes.

The veteran activists, when working the counter at Coffee Strong, present a similarly paradoxical image. The haircuts, physiques, and pieces of military uniform some sport communicate various things to different audiences, though they are all material statements of their identification as members of the larger military network. Physique, hairstyle, and clothing can be considered part of folk culture: people's manipulation, adornment, and other ways of presenting their bodies is culturally contingent, and these forms of expression, though they may be mass-produced, are often especially important in expressing group identification and differentiation. Within the Coffee Strong community, which includes both military people and civilians, veterans' personal presentation works to unify those with military experience and to differentiate them from those without.

The transgressive use of military paraphernalia is one way that these veterans express their rejection of authority and advocate for social equality and freedom of expression. As did veterans involved in the GI resistance movement during the Vietnam War, the Coffee Strong veterans play with military symbols to create messages that subvert their intended usage. Veterans wear items from their military uniforms in ways that do not conform to the military's policies and with clothing items that explicitly express opposition to the military. The military, of course, has strict policies about dress and hair, and the veterans are playing with these rules to criticize the military's strict control of its membership and to express their freedom from its grip.

They also play with juxtaposition. For example, during the period of my research, Joseph frequently wore his military camouflage blouse—the one he had worn in Iraq, with his authenticating nametag. His sculpted physique, short haircut, and wearing of military paraphernalia with his own name on it communicated his insider status. Yet, he wore the shirt unbuttoned and untucked, draped over Coffee Strong or other antiwar T-shirts. This recontextualizing of his military gear not only violated the military dress code but also semiotically placed the military symbol into dialogic interaction with overtly antimilitary T-shirts, thus creating that fruitful tension that Claude Lévi-Strauss called *bricolage*. Dick Hebdige explains bricolage as the combining of discrete communicative elements that "are capable of infinite extension because basic elements can be used in a variety of improvised combinations to generate new meanings within them," and that can be placed "in a symbolic ensemble" that serves "to erase or subvert their original straight meanings" (1979, 103, 104). These types of multivalent juxtapositions will be interpreted differently by different audience members, invoking anger in some and admiration or sympathy from others.

This has certainly been the case with Coffee Strong's "audiences." Because of many Americans' intense loyalty to the military and the emphasis on patriotism since the September 11 attacks, Coffee Strong veterans sometimes encounter hostility to their oppositional displays, and the coffeehouse has been vandalized on several occasions. The lack of a draft for the current wars also contributes to the negative attitude that some in the country have toward antimilitary activists. Because the current wars are being fought by volunteers rather than draftees, a number of people have told me that troops have no right to protest because they knowingly signed up and chose to make a commitment to the military. Once, when I was standing in the parking lot of a nearby restaurant filming protesters walking from Coffee Strong to a rally, the restaurant's owners came out and demanded than I get off their property. On another occasion, a bystander yelled obscenities, emphasizing specifically the lack of loyalty and patriotism of those marching in uniforms. Interestingly, however, Coffee Strong has received much less antagonism than its members anticipated, which they attribute to their strongly pro-soldier, though antiwar, stance.

The use of military paraphernalia by veterans involved in Coffee Strong is further complicated because some veterans' feelings about the military are fraught with ambivalence. Many chose to enter the military believing in its mission and proud of the opportunity to serve. Many come from families with histories of military service, and their friends and relatives are proud of their contributions to the country. Furthermore, many were drawn to Coffee Strong not for ideological reasons, but because, once in the military, they felt that they were mistreated or their needs were ignored. Some believe that war is necessary and identify with the military and military culture, but are critical of the internal functioning of the military or feel that the military is not adequately providing for its personnel. Others are not outright against war, but rather think that the current wars are unjust. Because of this diversity of perspectives, the wearing of bits and pieces of military garb can express combinations of pride in service, nostalgia for what they had hoped their experience would be, or ironic inversions. Kira, who was involved with Coffee Strong's advocacy for survivors of military sexual trauma, for example, explained that she was forced to leave the Air Force Academy after a fellow cadet raped her and her command persecuted her for reporting the crime. In her activism targeted at providing resources for other survivors, she always wore her combat boots, which she explained were an expression of her sadness at having been forced to dramatically shift the trajectory of her life. For Kira, wearing the boots expressed her ambivalence toward the military: an institution in which she had hoped to become a leader but that had traumatically disrupted her life.

OPPOSITIONAL POSITIONING, COMMUNITY, AND HEALING

Coffee Strong provides war veterans a space in which they can share tangled and often knotted feelings with a group that understands the emotional and intellectual struggles that they are dealing with. Many feel very strongly that the wars to which they contributed were unjust, and they deal with guilt about having participated in the disruption of people's lives, social injustices and, in some cases, killing. Most have memories of problematic, emotional, or traumatic moments that are difficult for them to come to terms with. Most have been officially diagnosed with PTSD, and even those who have not nevertheless struggle with various levels of combat stress.

Though it was created as an activist organization, one of the most important functions of Coffee Strong is to provide a space and community for its veteran members to cope and hopefully eventually heal. US Army soldier Shenandoah, for example, explained the role that the Coffee Strong community had for him: "Honestly, I think, groups like this, meeting you [points to Andrew, a veteran of the Marines and Seth, a veteran of the US Army], this place opening up, meeting you [points to Josh, a veteran soldier], and getting involved in things, honestly, I can say it saved my sanity. Because, I didn't have anybody to talk to. Because, I'm still active duty [pause]. I was honestly losing my mind until I met people who were out of the military or who were still in the military, but who were getting together outside the military to talk about these things." Because of the intensity of their experiences and because some of what happens in war is considered morally and psychologically reprehensible by many, some veterans feel comfortable sharing certain war memories only with other veterans. Though they have not shared most of these personal-experience narratives with me, an outsider, the friendships I established during the filming of the documentary did make me privy to some.

During the filming, we organized a group of veterans on December 17, 2009, to discuss their struggles with combat stress, which is when Shenandoah shared the above quote. Unlike the ego-bolstering stories of bravery and heroism that characterize some war recollections, those told at this session exposed the pain, ironies, injustices, and humor of war. Andrew explained to me in a phone conversation on September 2, 2011, that the activists refer to this type of anecdote sharing as the "race to the bottom," as tellers compete to describe why their experiences were worse than those of others. At this event organized for the filming, their interactions comprised the telling of anecdotes about their war experiences, their interactions with family and therapists, and their struggles and successes with depression, memories, and relationships. At one point, Andrew told about a get-together he had with members of his unit after they had returned from Iraq and shortly before they redeployed. In remembering this incident, he was reflecting on the humor and stories shared by his unit throughout its deployment and how far removed that talk was from what would be accepted as normal outside the war context. This gathering took place the evening before they were going to visit another member of his unit who had been injured. One of the Marines in his unit "lost his legs to an IED [improvised explosive device] that also blew up our company commander and one of our lieutenants, and another of our Marines. Or in English, killed three and severely wounded one." His unit mates had driven to where Andrew lived in Indiana, so that they could continue the next morning to Walter Reed Hospital to visit their injured friend. That evening, they went to a bar where they shared memories about their time in Iraq. Andrew remembered:

> There was the clapper, which was this dude who had been shot and was lying in the middle of the road, and the entire battalion basically had to drive over this guy to get where we were going because the road narrowed and went around a curve. So, we're talking about the clapper, because you drive over him and his upper body would flop up and his arms would flop up and he'd clap. So, we're telling this story, and my girlfriend at the time comes over, and is like, "Hey, what are you guys talking about?" [pause, nodding head] And I was like, "You probably really don't want to know what we're talking about right now." It was just the weirdest disconnect. As we're having this conversation, it's like the most normal conversation in the world, like you're talking about the weather. "Yeah, you remember that one dead guy?" And then the real world comes in, and you're just like, "Yeah, you probably don't want to hear about this."

After Andrew related this incident, Seth posed the rhetorical question, "How can you ever have a normal relationship with someone after you've been through all that?" Andrew responded, "Yeah, I'm still trying to figure that one out. I really, I don't know how." Seth joined in with "So am I," which was followed by mutterings of agreement across the room.

By telling this disturbing narrative to members of the insider group at Coffee Strong, Andrew provided an example of the inhumanity of war. He emphasized the cruel treatment of innocent Iraqis at the same time that he commented on the lack of humanity of the American troops in Iraq who laughed at this tragic incident. Since he was a protagonist in the story, though, he also referenced his own state of mind during his deployment, which was far removed from what he considered his "normal" self—someone who would never knowingly run over a dead body or laugh about doing so. As he told the story, those in his audience chuckled and nodded in support, remembering their own experiences and thus providing Andrew a nonjudgmental space in which to explore his own psychological contradictions. Sharing narratives that highlight the horrors of war and what young soldiers like Andrew perceive as assaults on their own humanity also helps to fuel the activism of other young veterans.

Andrew's story also gave the veterans the opportunity to reflect on how troops in need of catharsis use humor as a way to cope with trauma. In his examination of the humor that emerged after the 9/11 attacks, Bill Ellis explains that those most affected by the disaster, such as the emergency responders, found "themselves in an awkward bind, needing black humor to cope with the horror of the events but not being able to justify their actions to others" (2003, 37). Others have explored the use of humor by others whose professions bring them in regular contact with pain and suffering, for example, doctors (e.g., George and Dundes 1978; Moore 1991; Odean 1995; Gabbert and Salud 2009). Those dealing with the realities of war frequently use humor for a similar purpose.

Veterans sometimes choose not to share certain stories with outsiders because they fear being judged or misunderstood. For example, I have heard a veteran refer to the adrenaline rush and enjoyment he felt when exerting violence (including killing) on others. These emotions are difficult for many veterans to understand or admit to openly after they have (as they call it) "come down" to a less heightened emotional state—something that often does not occur until months after their deployments. It can be difficult for war veterans to accept this self-knowledge of how they felt and what they did. Like many veterans, those involved in Coffee Strong are kind and generous people who pride themselves on treating others with respect. How, then, can they come to terms with their memories of feeling joy at harming others? The Coffee Strong subgroup provides a space for them to explore these emotions in a less formal way than in therapy sessions. In relating these personal-experience narratives, they can share their conflicting feelings with others who are struggling with similar memories and who they expect will not judge them. Through participation in the community, they attempt to integrate these difficult memories into coherent narratives about their own identities (Linde 1993). Navy veteran Matt articulated this process especially well during the discussion about PTSD on December 17, 2009: "Coming here and meeting folks here has helped redefine the way I tell my story. Before, when people asked about being in the military, it was almost something like, I just kind of mumbled and stuttered and just kind of looked to the ground and just avoided it. But now, I tell my story from a whole different perspective."

The veterans also tell stories about moments that led to their transformations. Most have rehearsed narratives about their citizen/military/antiwar activist trajectory. They often include their motivations for joining the military, their perspectives on war and militarism when they first joined, key incidents that made them question the military, and what they are thinking and doing today. The frequent rehearsal of these narratives is important to their self-identification, but it is also necessary because as veteran antiwar activists, they are frequently asked questions by both those who are sympathetic and those who are critical about how they went from volunteering for military service to vociferously opposing the United States' current wars.

THE STRATEGIC USE OF OPPOSITIONAL POSITIONING IN ACTIVISM

The personal-experience narratives and anecdotes of veterans are especially effective elements of their political activism because they provide firsthand renditions of war. Following in the footsteps of GI resisters during the Vietnam War (Moser 1996; Fitzgerald 2007), veteran activists at the local and national levels have capitalized on the telling of their firsthand eyewitness testimonies as a political tool, for example, in the Winter Soldier testimonies organized by the organization Iraq Veterans against the Wars (e.g., Iraq Veterans against the War and Glantz 2008). Coffee Strong veterans speak at rallies, lectures, press conferences, and conventions, often sharing their experiences of war, the military and Veterans Affairs. Their stories simultaneously shock, raise awareness by putting a face to the realities of war, and invoke compassion for those who fight on the ground and are thus most impacted. This public sharing can evoke emotions and sympathy, which activists hope will draw people to their cause, and it disseminates information that they feel members of the general public should know. But it is important to note that these stories also generate anger and animosity from those with opposing political views, and thus simultaneously contribute to processes of differentiation.

The Coffee Strong veterans recognize the irony that their very engagement with that which they protest is what gives them access to speak and their voices legitimacy. At a rally in Portland, Oregon, on October 2, 2009, a number of civilian speakers provided statistics and information about America's current wars that were well received by an audience of mostly antiwar sympathizers. Then Coffee Strong founder Seth took the microphone and presented himself as a veteran of the Iraq War. The crowd became quiet, listened carefully, and cheered enthusiastically when he described his refusal to deploy to Afghanistan. The crowd seemed more affected by Seth's contribution to the rally than by the contributions of the nonveteran speakers. On the one hand, this made perfect sense. His participation in the war made him an especially significant antiwar advocate, and it was logical that members of an antiwar audience would celebrate someone whose politics had changed to resonate more with their own. On the other hand, I was struck by the irony that participation in the very thing that his audience opposed was what gave his words such credibility. When I raised this question with Seth after the rally, he responded: "Yeah, it's veterans' privilege. If I hadn't been to Iraq and done horrible things over there, I wouldn't be able to get up and talk in front of crowds. Whatever I had to say wouldn't be worth a damn. I could have some crazy ideas, but I went over to Iraq, so now I get to tell them to hundreds of people. I don't know, yeah, that is kind of an issue in our movement." Seth acknowledges the irony in this situation, while also recognizing that his special status as a warrior turned activist gives him a privilege and a responsibility to speak out as he did. "They've got to hear it from someone," he says, emphasizing that it is vital that people not involved in the war hear about its realities from someone who has had direct exposure.

Sharing personal-experience narratives about war and the military is also especially effective in the organization's networking with veterans and active-duty soldiers. In the day-to-day running of the coffeehouse, Coffee Strong veterans have determined that, in engaging with current troops about their military experience, it is not effective to accost them with antimilitary rhetoric. A better strategy is to draw on their firsthand esoteric knowledge about the dissatisfaction many military personnel feel about work conditions and relationships with their commands. The Coffee Strong folks often initiate conversations with active-duty customers by asking apolitical questions such as, "How is your command treating you?" This question often launches personal-experience narratives that the Coffee Strong veteran can follow up with his or her own. When the barista looks like a soldier and confirms this status with the use of enlistic speech, it facilitates shared identification and a first level of bonding. This often happens without overt political discussion coming into play. I have frequently observed that after exchanging a few stories, a customer leaves without any discussion of Coffee Strong's antiwar objectives, and without any attempt on the barista's part to move the interaction toward the organization's goals.

Nonetheless, because these conversations happen in a space adorned with antiwar décor, the politics of differentiation is also in play. Customers have varying degrees of awareness of this differentiation, and of the political nature of the coffee shop, but it is hard for any of them to ignore the pervasive political messages and frequent politically charged conversations

9.2. Coffee Strong members at an antiwar demonstration.

occurring in the background. As a result, military customers sometimes make jokes or comments about the irony of their being there. At the same time, these informal story-laden conversations are an important means by which Coffee Strong makes itself known as a nonconfrontational space, for active-duty personnel as well as for activists. The informal, nonpolitical conversations welcome newcomers into the shop, letting them know that other veterans are available to chat there and offering opportunities to bond with members of the Coffee Strong community. These interactions can eventually lead to more active engagement or use of the organization's resources in future visits.

Some Coffee Strong veteran activists also strategically capitalize on their oppositional positioning in their protests (figure 9.2).

An overpass that provides access to the bases crosses over Interstate 5, linking the gate into the base, east of the highway, with the town of Lakewood and Camp Murray, a Washington National Guard base, on the west side. This bridge, about 300 meters from Coffee Strong, is a site for regular displays of pro-military patriotism by people waving American flags and "Support our Troops" signs. Coffee Strong also uses this bridge for its rallies, but its demonstrations advocate for soldiers' rights and protest against specific military policies or actions. In much of the activism and discourse surrounding the United States' current wars in Iraq and Afghanistan, a sharp divide is made between those in the military who are presumed to support the war, military, and government, and those who are protesting the wars, who are presumed to be outside the military. When they wear uniforms at these public events, Coffee Strong veterans capitalize on their membership in the military network in a variety of ways. The majority of the people crossing the bridge are people coming on and off the bases. The uniformed Coffee Strong members are thus simultaneously protesting the military and connecting with active-duty people, letting them (and civilians) know that there is a voice of resistance within the military. This is especially significant to them because they all experienced alienation both during their military service and after, and they are acutely aware that other active-duty personnel feel similarly alienated. Part of Coffee Strong's activism is to let soldiers and veterans who feel this way know that they are not alone.

The Coffee Strong veterans direct their activism near the bases to specific issues rather than focusing on general antiwar messages. They consciously do not promote the use of the peace sign in their demonstrations because they think it will alienate most soldiers and does not adequately represent their objectives, though often civilian peace activists in their demonstrations do showcase peace signs and other antiwar symbols (figure 9.3).

Rather than promoting what they perceive to be the unrealistic goal of achieving world peace, they are committed to improving the situation for soldiers and providing resources for them, and thus they target what they believe to be injustices within the military, for example, the treatment of military prisoners, the "stop-loss" policy of extending enlistment commitments without soldiers' consent, the failure to aggressively punish perpetrators of sexual abuse, and the lack of adequate medical resources for those suffering from post-traumatic stress and other common health problems. They hope that their signs at protests, addressing issues that many servicemen and servicewomen can relate to directly, combined with the wearing of their own military uniforms, will communicate to the military that veterans are not afraid to speak out about their experiences. At a rally in support of two soldiers imprisoned at Fort Lewis on October 18, 2009, I asked a former US Marine who prefers to remain anonymous why he wore his uniform, and he answered: "I'm wearing my uniform because I want to show the public that it's not just civilians that are here today supporting these two soldiers. Veterans are also here. Veterans of the Iraq War, Afghanistan War, and also veterans who have resisted war themselves. By wearing my

9.3. Antiwar demonstration, with peace sign.

uniform, I'm hoping to show that there are veterans here today that are showing support for [the prisoners] in the hopes that Fort Lewis at least honors their civil rights." Other messages also come into play in this example. This Marine suffered from injuries during his service in Iraq that leave him struggling with severe physical and psychological problems resulting from a traumatic brain injury (TBI), PTSD, and physical impairment. His physical appearance at the rally contrasted markedly with the conventional image of a strong, healthy young Marine. His hair was disheveled; his mind was disoriented from the numerous medications he was taking for epilepsy, depression, and schizoaffective disorder; and he walked with a cane. His participation was a real-life presentation of what these wars are doing to some of the people fighting them. At the time of these rallies, he was struggling with Veterans Affairs to get the medical care he needed, so his participation also communicated information about the inadequate care he was receiving from the government bureaucracy. His participation also worked to draw sympathy from would-be and current activists, as he put a human face to the trauma suffered by many of those serving in these wars.

CONCLUSION

Exploring the antiwar activism of veterans may seem like a paradoxical contribution to a volume about military folklore. However, understanding the ways in which the Coffee Strong veteran activists engage with military folklore is important because it yields insight into an often-neglected subculture of the US military. Even though these young veterans are "oppositionally positioned" toward the military establishment, they are still very much a part of the military culture, and their interactions with other veterans and active-duty personnel suggest a psychological complexity in that culture that official brochures and publicity may not reveal.

All wars and all militaries involve participants who are continually thinking, feeling, questioning, and wondering about the meaning of what they are doing and whether the goals of their command justify their actions and sacrifices. In the current wars in Iraq and Afghanistan, many American troops would answer yes, while others would answer no. Regardless of how they answer this question or how they position themselves politically—and it is important to recognize that within the military are people across the political spectrum—everyone experiences war with ambivalence. It is not a game; troops have to do difficult things; and some of what they do and see does not fit their own sense of morality. Furthermore, the military is a hierarchal institution with few avenues for those in its lowest strata to express their opinions or criticism. Unlike the case with most jobs that Americans have when they are young—most of the Coffee Strong veterans joined the military straight out of high school—once someone signs up, he or she cannot legally quit until the military allows it. As with any hierarchy that one cannot leave, some of those at the bottom will feel mistreated. The many conversations I have had with active-duty and veteran military personnel as part of this and another project reveal that the identification and differentiation shared by Coffee Strong members exist among others in the military, even if their "oppositional positioning" is not always as politically motivated as it is for Coffee Strong veterans (Gilman 2010, forthcoming). This is a topic for further research.

Their military experience impacted the Coffee Strong veterans profoundly and is shaping their lives as they create futures for themselves. Though these veterans are striving to develop their identities explicitly apart from the military, many also do not want to mute or stifle the memories of their military experience, even the guilt-laden or otherwise painful ones. Though some wish that they could be freed from such memories or turn back the clock and make different decisions, others feel that their war experiences provided them with valuable insights that they would not otherwise have had. Telling their stories—including the painful ones—while participating in other military-related folklore is therefore a necessary element of their self-definition.

They also see it as a political obligation. Some express frustration at the apathy of many Americans about the wars, and they feel that it is only by having to face the truths of soldiers' experiences directly that people will be motivated to effect change. They recognize this possibility in themselves, and so by telling their stories and continuing to associate themselves with the military through the use of symbolism and other expressive culture, they force their audiences to engage with the realities of war while at the same time keeping their memories alive.

WORKS CITED

Bauman, Richard. 2000 [1972]. "Differential Identity and the Social Base of Folklore." In *Toward New Perspectives in Folklore*, ed. Américo Paredes and Richard Bauman, 31–41. Bloomington, IN: Trickster Press.

Burke, Carol. 1989. "Marching to Vietnam." *Journal of American Folklore* 102 (406): 424–41. http://dx.doi.org/10.2307/541782.

Cain, Carole. 1991. "Personal Stories: Identity Acquisition and Self-Understanding in Alcoholics Anonymous." *Ethos* (Berkeley, Calif.) 19 (2): 210–53. http://dx.doi.org/10.1525/eth.1991.19.2.02a00040.

Cortright, David. 1975. *Soldiers in Revolt: The American Military Today*. New York: Anchor Press.

Ellis, Bill. 2003. "Making the Big Apple Crumble: The Role of Humor in Constructing a Global Response to Disaster." In *Of Corpse: Death and Humor in Folklore and Popular Culture*, ed. Peter Narváez, 35–79. Logan: Utah State University Press.

Fine, Gary Alan. 1979. "Small Groups and Culture Creation: The Idioculture of Little League Baseball Teams." *American Sociological Review* 44 (5): 733–45. http://dx.doi.org/10.2307/2094525.

Fitzgerald, John J. 2007. "The Winter Soldier Hearings." *Radical History Review* 97: 118–22. http://dx.doi.org/10.1215/01636545-2006-017.

Gabbert, Lisa, and Anton Salud, MD. 2009. "On Slanderous Words and Bodies Out-of-Control: Hospital Humor and the Medical Carnivalesque." In *The Body in Medical Culture*, ed. Elizabeth Klaver, 209–27. New York: SUNY Press.

George, Victoria, and Alan Dundes. 1978. "The Gomer: A Figure of American Hospital Folk Speech." *Journal of American Folklore* 91 (359): 568–81. http://dx.doi.org/10.2307/539575.

Gilman, Lisa. 2010. "An American Soldier's iPod: Layers of Identity and Situated Listening in Iraq." *Music and Politics* 6 (2).

Gilman, Lisa. 2011. *Grounds for Resistance*. Film.

Gilman, Lisa. Forthcoming. "Music and the Ambivalence of War for American Troops Fighting in Operation Iraqi Freedom." In *Post-Conflict Music: Global Rhythms of Resistance*, ed. Colin Wright and Lucio Spaziante. Nottingham: Critical, Cultural and Communications Press.

Hebdige, Dick. 1979. *Subculture: The Meaning of Style*. London: Methuen Press. http://dx.doi.org/10.4324/9780203139943.

Iraq Veterans against the War and Aaron Glantz. 2008. *Winter Soldier: Iraq and Afghanistan Eyewitness Accounts of the Occupations*. Chicago: Haymarket Books.

Linde, Charlotte. 1993. *Life Story: The Creation of Coherence*. New York: Oxford University Press.

Moore, Jamie. 1991. "Poetry, Puns, and Pediatrics: The Verbal Artistry of Dr. James L. Hughes." *North Carolina Folklore Journal* 38 (1): 45–71.

Moser, Richard R. 1996. *The New Winter Soldiers: GI and Veteran Dissent during the Vietnam War*. New Brunswick, NJ: Rutgers University Press.

Nicosia, Gerald. 2001. *Home to War: A History of Vietnam Veterans' Movement*. New York: Crown Publishers.

Norris, Robert. W. 2008. "A Comparison of American GI Resistance to the Vietnam War and the Iraq War." *Fukuoka International University Bulletin* 20: 1–14.

Noyes, Dorothy. 2003. "Group." In *Eight Words for the Study of Express Culture*, ed. Burt Feintuch, 7–41. Urbana: University of Illinois Press.

Odean, Kathleen. 1995. "Anal Folklore in the Medical World." In *Folklore Interpreted: Essays in Honor of Alan Dundes*, ed. Regina Bendix and Rosemary Lévy Zumwalt, 137–52. New York: Garland Publishing.

Part IV
Remembering

10

Colonel Bogey's March through Folk and Popular Culture

Greg Kelley

Frederick Joseph Ricketts, venerated as the "British March King," stands as one of Britain's finest composers of military music. He is touted as England's answer to Sousa. Born in East London to a Shadwell coal merchant in 1881, Ricketts was orphaned by the age of fourteen and lied about his age in order to join the army as a bandboy. In 1904, after seven years' service in India, he returned to England to study at the Royal Military School of Music where, surprisingly, he was ranked bottom of the class for his original march composition. In 1908 he took charge of his first band, but with the onset of World War I his bandsmen were called back to service and Ricketts found himself directing a band composed primarily of overage musicians and bandboys. In that capacity, he spent the war leading concerts for service charities and other wartime causes (Richards 2001, 428–29). After the war, Ricketts directed several military bands, transferring finally to the Divisional band at Plymouth in 1930, where he served as director of music for the Royal Marines until his retirement as a major in 1944. He was composing all the while, using a pseudonym (Kenneth J. Alford) because service personnel during this time were discouraged from developing professional careers outside the military. Of the eighteen marches he composed in the span of his active career, the best known by far is "The Colonel Bogey March," familiar to most Americans as the theme melody from the film *Bridge on the River Kwai*.

Alford composed "Colonel Bogey" in 1914, a portentous year for marches, to be sure. Legend has it that the original inspiration for the march

DOI: 10.7330/9780874219043.c10   205

came from an eccentric colonel whom Alford met while stationed at Fort George near Inverness in Scotland just before the war began. The composer played golf, and on the local course he occasionally encountered the colonel, whose nickname among fellow golfers was "Bogey." Instead of shouting "fore" to warn of an impending drive, the colonel had the peculiar habit of whistling a two-note phrase, descending in the minor third. This little musical figure took root in Alford's receptive mind—spawning the signature motif of his memorable march.

When Alford penned "Colonel Bogey," he could not have foreseen its destined fame. One authority on the bawdy folk culture of WWI, Brophy and Partridge's *The Long Trail*, claims that "Colonel Bogey" was "the most frequently heard marching tune in the Army . . . whistled and hummed everywhere" (1965, 16). It was to be Alford's masterwork and a significant commercial success. By the early 1930s, the sheet music of the march had sold more than a million copies, and the tune had been recorded countless times (Graves 1999, n.p.). Beyond licensing and recording, however, "Colonel Bogey" enjoyed a vibrant other life in parody.

EARLY PARODIES

Ultimately, what permanently stamped the melody into the popular consciousness was not its commercial achievement but rather its function as a vehicle for numerous comical folk lyrics. As Jeffrey Richards observed, "['Colonel Bogey's'] rhythmic structure was so appropriate for words that marching soldiers rapidly attached to it obscene lyrics, which were cheerfully sung by squaddies [British Infantry soldiers] in both world wars" (2001, 431). The catchy two-note introductory figure lends itself to bawdy disyllabic lyrics, and within a few years of its composition, the march had inspired several ditties like this one:

> Bullshit! That's all the band could play,
> Bullshit! They played it ev'ry day
> Bullshit! Ta-ra-ra bullshit!
> Ta-ra-ra bullshit! bullshit! bullshit!

Given that Alford's marches were composed as "affirmations of his patriotism," "tributes to the fighting forces," and "morale boosters" (Richards 2001, 431), this stanza appears lyrically hostile to those intentions. The band is roundly berated, and then, suggestively, patriotism is impugned as well. "Ta-ra-ra bullshit" unmistakably invokes "Ta-ra-ra-boom-de-ay," the song made famous by Lottie Collins, the British music hall singer who first performed it in London in the 1890s. With an accompanying dance routine, she performed the song all over London and then toured America and Australia. The song was memorialized around the world and by WWI had become shorthand for British patriotism ("A Chat" 1895; Busby 1976, 39). So, with their lyrics, British soldiers were simultaneously undermining the understood jingoism of both "Ta-ra-ra-boom-de-ay" and "Colonel Bogey." Military hardliners may have reacted like WWI British commander in chief Douglas Haig upon overhearing one battalion singing a bawdy rendition of "Turkey in the Straw": "I like the tune . . . but you must know that in any circumstances the words are inexcusable" (Macdonald 1993, 208).

Poetically, however, this lyric to "Colonel Bogey" is not especially imaginative, nor does the last line scan easily with the original rhythm of the march. We might compare another rendition from the same period, whose irreverence, stated in terms of the disintegrated body and cannibalism, is considerably more graphic:

Bollocks, and the same to you,
Bollocks, they make a damned good stew,
Bollocks, mixed up with scallops,
And a nice tasty arsehole or two.

(Sometimes an alternate third line is inserted: "Knackers, go well with crackers.")

Considering the parodies that would imminently emerge during WWII, it is notable that by the mid-1930s the word bollocks figured prominently in many British renditions of the march: in the folk mind the melody was already being associated with the realm of the testicular. Although Alford was dismayed at the lyrics that had become attached to his composition (Trendell 1991, 33), on occasion he affably recounted an incident that had occurred in 1919, while he was listening to the Royal Marine Artillery Band playing at Ryde, Isle of Wight. "As the band struck up 'Colonel Bogey,' an unknown officer turned to [Alford], who was dressed in civilian clothes, and said 'There's that bollocks tune again! How I'd like to strangle the bloke that wrote it'" (quoted in Trendell 1991, 67).

WORLD WAR II AND THE THIRD REICH

From there it was a logical progression that led to the most persistent—and memorable—adaptation of the march, "Hitler Has Only Got One Ball," which appeared initially among British troops sometime in 1939 and remains in oral tradition even today. One typical variant goes this way:

> Hitler has only got one ball,
> Göring has two but very small,
> Himmler is rather sim'lar,
> But poor old Goebbels [Go-balls] has no balls at all.

A means of ridiculing the Nazis, "Hitler Has Only Got One Ball" became immensely popular among both British and American troops (Cleveland 1994a, 85), who in transmitting this song were exercising something of a wartime convention by demeaning the sexual faculties of enemy leaders. But the mockery extended beyond just the Nazis' sexual capacities. Since the 1920s, the words *balls* or *ballsy* had come to denote notions of courage, nerve, or fortitude. In that sense, defective testicles rendered the Nazis defective soldiers. This song's itemized taxonomy of malformed German genitalia—the monorchid, the micro-orchid, the anorchid—was particularly forceful, and satisfying, to Allied soldiers in that it scattered satiric buckshot across the whole Nazi high command (Hitler; Hermann Göring, commander in chief of the Luftwaffe; Heinrich Himmler, Reichsführer of the SS; and Goebbels, Reich minister of propaganda).

The song's genesis in this form is unclear. In his autobiography *Fringe Benefits* (2000), Anglo-Irish writer Donough O'Brien claims that the original was written by his father, Toby O'Brien, in August 1939 when the latter was working as a publicist for the British Council. His version started with the words "Göring has only got one ball," and went on to describe Hitler's two diminutive ones—contradicting virtually all later versions, in which the positions are transposed. But the composer and broadcaster Hubert Gregg also professed to have authored the lyrics. Allegedly, he penned the bawdy verse and proffered it anonymously to the British War Office, which welcomed the contribution enthusiastically as an implement of wartime propaganda. The award-winning BBC radio play by Neville Smith titled *Dear Dr. Goebbels* (2001) posits yet another origin of the song. Broadcast in 2000, the play, purportedly nonfiction, recounts the story of a WWII prosthetic seller named Philip Morgenstern who goes undercover with MI6 to Germany. There, he meets Goebbels and is commissioned to fit the doctor with new prosthetic surgical boots. In the process, he learns scandalous details of Goebbels's private life, which he then versifies to the tune of the "Colonel Bogey March." Churchill later offers Morgenstern an honor, officially for bravery in service but unofficially for the propagandistic musical jewel. The play closes with Churchill, on his last day of office, singing the song with Morgenstern. These claims all remain unsubstantiated, and authorship of "Hitler Has Only Got One Ball" has never been definitively established. There is no known attempt by anyone to acquire or enforce a copyright on the lyrics—listed in the Roud Folk Song Index as number 10,493. In 1939, Ray Sonin set wholesome patriotic lyrics to Alford's march with "Good Luck and the Same to You," but the sanitized song failed to obscure the bawdy folk versions already in circulation.

It is impossible for us to know just from the texts what the verses may have meant to specific informants at the time, but it is probable that this song would have elicited from Allied soldiers some amusement at disparaging the enemy with the added social benefit of morale boosting. It would be misleading, though, to suggest that the song served an express institutional strategy of propaganda. As Christie Davies has argued, political and military institutions are generally unsuccessful in using humor as a strategy in war, strictly speaking. "Jokes belong to the black market" rather than to authorities, and "popular humour operates independently of any strategy that the official controllers of war-time propaganda would like to impose on it" (Davies 2001, 402). The strategic prospects for authorities are doubly limited when the humor is conveyed through melody, because a song's success, like a joke, depends upon audience reception rather than political imperative. According to Les Cleveland, a noted scholar of military occupational lore, the US music industry's efforts to establish a "correct" formula for WWII songs were ineffectual. The U.S Office of War Information considered the potential of popular songs as military propaganda, only to determine that there "was no payoff in high purpose or patriotic intent. A song had to be accepted by the mass audience on its own merits" (Cleveland 1994b, 167). So it was with "Hitler Has Only Got One Ball," which obviously found fertile soil into the 1940s but has flourished in folk and popular culture ever since—beyond the military, beyond the war—notwithstanding (dubious) assertions, like O'Brien's, that it was launched as a propagandistic initiative through government channels.

Some propose that the song, authored or not, developed from a simple linguistic glitch—namely, the difficulty of English speakers to pronounce

Goebbels's name properly—and that the rest of the song unfolded organically from there. It is worth noting that some later variants among children, who probably had no idea who Goebbels was (and who probably could not pronounce his name), replaced "Go Balls" with a fictional character called "Joe Balls." But regardless of the subject's name, all collected versions of the opening stanza reach the same poetic denouement at line 4: "no balls at all." That is to say, in this musical catalog of testicular disorders, the definitive last entry is always anorchism—the physical signifier of a lack of courage or character.

A second, less familiar, verse adds more details about other Third Reich personalities and introduces the notion of deformity in terms of overabundance rather than deficiency:

> Rommel has 4 or 5, I guess,
> No one's quite sure bout Rudolf Hess,
> Schmeling is always yelling,
> But poor old Goebbels [Go-balls] has no balls at all.

Unlike his counterparts, [Erwin] Rommel's affliction is one of excess. If in these rhymes testicles equate with fortitude, then Rommel, Germany's highly decorated field marshal, comes across as hypermasculine, which diverges from the general tenor of the rest of the song. He does not fit easily into the song's central message because he was the least hated of the Nazi high command: a respected adversary who was one of the few senior officers not involved in war crimes, the "Desert Fox" was even implicated in the plot to overthrow the Führer near the end of the war. These ambiguities may account for the relative obscurity of this verse compared to the well-worn first verse. It is telling that Goebbels, who was among the most reviled Nazis, appears in virtually every variant and suffers most severely in the song's inventory of testicular defects. As for Rudolph Hess, Hitler's devoted deputy Führer, no one is sure about him, perhaps because of the mysterious circumstances surrounding his capture and incarceration. He remains one of the most enigmatic personalities of the Third Reich: he secretly flew solo to Scotland in 1941 to negotiate peace with England but was arrested and detained for the remainder of the war and then imprisoned for life after a conviction at the Nuremberg trials.

The appearance of the boxer Max Schmeling in line 3 may strike us as more puzzling yet. As the only nonsoldier and the only one without an expressed testicular abnormality, he seems at first to be incongruously yoked with the others. But any American GI old enough to fight in WWII would well remember Schmeling's backstory. His matches against Joe Louis were monumental events of 1930s. Schmeling delivered the best performance of his career in 1936, knocking out in the twelfth round the then undefeated Louis. Back in Germany, Schmeling was swept up in the Nazi propaganda machine and his victory was trumpeted as evidence of Aryan superiority. The subsequent rematch between Schmeling and Louis for the World Championship in 1938 at Yankee Stadium was at that time the most anticipated and politicized sporting event ever on American soil. Billed as the "Battle of the Century," it had deep racial implications and became an epic morality play between the forces of Nazism and American democracy. Louis beat Schmeling soundly with a technical knockout in the first round. A decisive punch of the match was a blow to the left kidney, at which point Schmeling "grimaced and let out a high pitched cry that echoed throughout the stadium" (Margolick 2005, 298). Symbolically, this was a yell heard around the world; countless spectators and sports writers commented on the scream and it was remembered vividly as part of the national narrative about the fight. And within five years of the match, we find American versions of "Hitler Has Only Got One Ball" with the Schmeling line inserted. "Schmeling is always yelling" is an irresistible rhyme, but it clearly is there for more than just poetic reasons. It is an iconic moment of American victory over Nazism frozen in time. In the artistic frame of the verse, Schmeling is always yelling; that is, he is doomed to relive the moment of defeat perpetually. That image of the enemy would serve American soldiers well, and it fits precisely with the song's theme of derision.

Other subsequent British verses amended Hitler's bizarre testicular history, as it were, and narrowed the audience appeal through regional vocabulary and reference to local landmarks:

Hitler has only got one ball,
The other is in the Albert Hall,
His mother, the dirty bugger,
Chopped it off when Hitler was small.

She threw it, into the apple tree
The wind blew it into the deep blue sea
Where the fishes got out their dishes
And ate scallops and bollocks for tea.

It's not entirely implausible that Hitler may have been monorchid (as we shall see), or that his mother may have been abusive to the point of mutilation (an explanation for his pathology, perhaps). But it is unfathomable how, in the "plot" of the song, his severed testicle could find its way to such a place as the Royal Albert Hall after (nonsensically) having been a side dish for anthropomorphized fish at high tea. Founded in 1871, the Albert Hall is a famous performance and exhibition venue in Westminster and appears most often in these additional stanzas. It is significant that the ultimate resting place of the Führer's testicle is an arena of public display. Exhibition in that kind of space is a public assertion of ownership and subjection. The Albert Hall, then, becomes the municipal custodian of a wartime trophy, a symbolic means by which the British state is taking proprietary rights of Hitler's diminished power. Variants from other regions of the UK provide a case book of oikotypes, as the second line of the stanza is frequently altered to reference local buildings. In Manchester the other [ball] is in the "Free Trade Hall"; further north it is the "Leeds Town Hall"; in Glasgow the "Kelvin Hall"; and in Northern Ireland, the "Ulster Hall." As a narrative device, this localization emphasizes the singer's implied "civic" participation in immortalizing and exhibiting domination over Hitler.

Despite the poetic function of celebrating Nazi testicular deformity, some individuals have sought empirical explanations for these lyrics. An alleged Soviet autopsy on Hitler's remains made shortly after the war, released in 1968, is clear in its assertion of monorchism (see Bezymenski 1968; Waite 1977, 150). But the medical report has been dismissed by historians as its own bit of propaganda (see, e.g., Ainsztein 1969; Rosenbaum 1998). The authenticity of the autopsy is questionable, given that Hitler's death by suicide and the subsequent almost complete burning of his body left little for doctors to examine. Records do show, however, that as a soldier in the German Army during World War I Hitler was wounded in 1916 during the battle of the Somme, and although sources disagree as to the exact location of the wound, a few maintain that it was in the thigh or the groin. Some historians postulate that Hitler may have been monorchic from birth or that one of his testicles may have failed to descend at puberty. In any case, Hitler's World War I company commander claimed that a VD exam revealed that Hitler had only one testicle (letter to *Die Zeit*, December 21, 1971, cited in Waite 1977, 152, 450n65). "Psycho-historians" with a Freudian bent have made much of these findings, viewing Hitler's "putatively half-empty scrotal sack as the root cause of his murderous character, his sexuality, and his anti-Semitism. The rumor offers one-stop shopping for Hitler explainers" (Rosenbaum 2008, n.p.). For example, Robert G.L. Waite, whose book *The Psychopathic God: Adolf Hitler*

(1977) is often touted as the definitive psychological portrait of Hitler, argues that Hitler's alleged childhood monorchism formed in him a lifelong complex of deficiency, which contributed significantly to his pathological personality. "The problem of the Fuehrer's testicles" is a foundational premise of Waite's biography, and he declares outright that "the British Tommies were right all along in the first line of their version of the Colonel Bogey March" (although, he adds, "they were manifestly mistaken in the last [regarding the supposed anorchism of Goebbels, who fathered six children]") (1977, 150).

In November 2008, the discovery of an eyewitness account by World War I army medic Johan Jambor electrified the press. Jambor was said to have discovered an injured Hitler at the battle of Somme in 1916 and saved his life. Hitler was screaming for help as a result of injuries. Apparently, in the 1960s Jambor had revealed all this to a priest, who then dutifully wrote down the revelation in a document that surfaced in 2008, twenty-three years after Jambor's death (Roberts 2008, n.p.). Appearing originally in the *Sun* (London), the story was quickly picked up by other Commonwealth and American tabloid newspapers, which had a field day with the headlines: "One Ball, After All," "Half the Man He Used to Be," "Ballsy Ditties," and this favorite from the online newspaper the *Register*: "Hitler Had One Ball: Official—Other Not in Albert Hall, However." These headlines demonstrate that the lyrics were still well known in 2008. A separate, less publicized account says that his genitals were mangled by a goat when, as a child, Hitler tried to urinate into its mouth (Redlich 1999, 18).

In any case, the proximate cause of Hitler's presumed disorder (be it birth defect, bombshell, or goat) and whether it sprung from deliberately publicized misinformation or verifiable historical fact are not the central concerns. What irresistibly attracted the popular consciousness of the 1940s was the very idea that Hitler and his eugenic minions would have defective physiognomies. That notion, coupled with the medium of an already famous song, had all the makings of what we might call a "folk hit." And "Hitler Has Only Got One Ball" has remained on the charts ever since.

THIRD-WAVE POPULARITY: POST-WWII

Assorted postwar variants of the first line delineate Hitler's peculiar testicular "arrangement": "only one big ball"; "only one left ball" (a curious description, given that having one left testicle is considered the norm); "only one meat ball" (certainly suggestive of the testicular, but specific to the mid-1940s as it emerged just after the Tin Pan Alley song "One Meat Ball"—a novelty tune about the high cost of eating out—had been made famous by the Andrews Sisters).

Any chance of the melody slipping from popular awareness evaporated when the tune appeared memorably in David Lean's Oscar-winning *Bridge on the River Kwai* in 1957. In fact, the march is often referred to simply as the *River Kwai* theme, with the common assumption that it was composed specifically for the film. But the march's rich prior folk history unquestionably figured into Lean's decision to place it in the film. He remembered the parodies from his youth, and he originally wanted the British soldiers in the film to be singing the lyrics of "Bollocks, and the same to you" or "Hitler Has Only Got One Ball" as they entered the camp (Phillips 2006, 236). Producer Sam Spiegel, however, found the lyrics too offensive and Alford's surviving widow granted permission to use the melody only under the express condition that the bawdy lyrics not be included. It is reported that she cautioned the producers unequivocally, "A lot of rude words have been made up around that song and I don't want my husband made a mockery of" (Brownlow 1996, 354). Thus, the now-famous whistling version was substituted. Malcolm Arnold, the film's music director, recalled the method of his musical arrangement: "The whistlers. . . were a piccolo and seventeen members of the Irish Guards. They weren't handpicked; anybody can whistle. I said, 'Look, gentlemen, we all know both world war versions of "Colonel Bogey." But here, because of censorship, you've got to whistle it'" (Brownlow 1996, 381). In the end, Lean was pleased with the whistling rendition, conceding that even without the bawdy lyrics, "the English audience will know what we're after" (Brownlow 1996, 351). By that, he presumably meant the yoking together of martial protocol with implied irreverence. To the degree that Lean was correct about the audience's perception is a testament to the general awareness of "Colonel Bogey" parodies in Britain. Since then, the number of pop culture allusions, citations, oblique references, and send-ups of the song has been remarkable.

On the *Benny Hill Show*, for instance, the melody of the "Colonel Bogey March" was used in a sketch (original air date, April 16, 1980) mimicking the game show *Name that Tune*. Asked to identify the song, one female contestant guesses "After the Ball," which the host (played by Hill) tells her is half right. Then the gentleman contestant guesses "The Cobbler's Song," to uproarious laughter from the audience (*cobblers* being British slang for *testicles*). His second guess is the 1944 novelty song "I've Got a Lovely Bunch of Coconuts." Much is left unsaid here, but the allusion is transparent. Associating the melody of the "Colonel Bogey March" with anything ballistic instantly conjures up the song about Hitler.

Similarly, the lyrics are alluded to in a 2003 advertisement for an English ale called Spitfire. A banner caption reading "Spot the ball" accompanies a photograph of Hitler in full Wehrmacht uniform. The advertisement refers to print media "Spot the ball" promotions that used soccer match photographs from which the ball had been electronically edited out. Readers would then compete by guessing the position of the missing ball. The phrase "Spot the ball" is innocuous by itself, but when paired with the photograph of Hitler the two together become unmistakable imagistic shorthand of the theme that has been propagated by the folksong for more than seventy years. This is reminiscent of a truncated form that Bill Ellis has termed legend metonym, which reduces narrative to the level of simple allusion. These metonymic forms emerge when texts are so familiar that audiences "no longer need to re-experience them through performance"; that is, the well-known texts are distilled down to abbreviated verbal tags, such that they become "a name in form as well as function" (1989, 40). In the same way, the beer ad and Benny Hill's sketch work comically only if the fuller source text of "Hitler Has Only Got One Ball" is well known (which it most certainly is).

In his 2007 comedy tour *Fame*, Ricky Gervais builds a full minute of material around the song, emphasizing the endurance of the "Albert Hall" stanza and the logical problematics of it:

> No, it's a great gig, and it is at the Albert Hall, home to Hitler's other testicle. [laughter] "Hitler has only got one ball / The other is in the Albert Hall."
> I didn't see it, and I looked around for it, which is suspicious. If I had that, I'd have it in the foyer, on a plinth in a glass cabinet . . . in an eggcup. [laughter] I'd love to look at that, not in a gay way. I wouldn't be going, "Oh, there's a lovely bit of old bollock." I'd be going, "That's the physical and symbolic embodiment of pure evil. That's Hitler's seed."
> "Hitler has only got one ball / The other is in the Albert Hall / His mother has got the other." What? Did he have three? It doesn't make any sense at all. [laughter] Why would she have that? Why would your mother keep that, really? Unless she's going "Look, he's my son, and I love him. He's a wrong 'un. I know it. I don't want him breeding. I keeping back Hitler's testicle." Well, she would say "my Adolf's," wouldn't she? That name's died out, hasn't it? [laughter]

These tenacious folk and popular adaptations of "Hitler Has Only Got One Ball" have engendered what Hitler biographer Alan Bullock labeled disdainfully as the "one ball business" (see Rosenbaum 1998, 78). But it should not surprise us, given its immense popularity, that the melody of "Colonel Bogey" inspired a number of other folksongs as well.

CHILDREN'S APPROPRIATIONS

The march has become a classic in children's folklore, existing in multiple versions that mark it as a vibrant artifact of post-WWII oral culture. Most Americans who were children during the 1960s and 1970s will likely remember the playground songs about Comet cleanser, which took "Colonel Bogey" into the world of consumer goods.

> Comet, it tastes like gasoline,
> Comet, it makes your teeth turn green,
> Comet, it makes you vomit,
> So get some Comet and vomit today.

Children's folklore draws a wealth of material from commercial culture, and popular advertisements are frequently adapted and satirized on the playground. Some folk parodies of commercials outlast by decades the advertising campaigns that initially inspired them. For example, children still sing spoofs of "McDonald's Is Your kind of Place" commercials, which originally aired in 1967 and ran for only a few years after that. Similarly, variants of "Comet" linger in oral tradition more than forty years since it first appeared in the late 1960s. The song is not a parody, technically, because there was no existing jingle from Comet commercials that children then adapted and mocked. Rather, like the singing soldiers before them, they crafted new comical lyrics for music that was already floating in the popular ether. "Colonel Bogey," by then undergoing a revival following *Bridge on the River Kwai*, rhythmically accommodated the disyllabic word "Comet" perfectly (as it had done earlier with the names of Hitler and his generals). And because the product's name formed such a conspicuous rhyme with the word "vomit," most variants predictably became fixated on that bodily function. Indeed, like the standard Goebbels line in wartime versions, in the many incarnations of this folksong the "conclusion" of line 4 remains stable: the listener is invited to ingest the cleanser and enjoy its emetic properties. There is wide variance in the first two lines, however, where the rhyming possibilities seem endless—as in these few examples:

> Comet tastes like Listerine,
> Comet will make your eyes turn green.
>
> Comet, it makes your teeth turn red,
> Comet, it makes you wet your bed.
>
> Comet will get your bathroom clean,
> Comet will make your hair turn green.
>
> Comet, it makes your mouth turn square,
> Comet, it makes your butt grow hair.

Sherman and Weisskopf posit that rhymes about cleaning products appear in children's lore as a subversive response to familiar parental warnings about handling and ingesting such substances (1995, 162). But advertising itself is also targeted here. Constantly assailed by commercial images and jingles, children grow to resent the bombardment and use folklore as a means to "defend themselves against becoming mere bundles of reflexes that are dominated by the ads" (Knapp and Knapp 1976, 165). With songs like "Comet," they scoff at the relentless advice that advertisements have foisted upon them. Moreover, the song suggestively mocks commercially projected claims about domestic perfection. Uncleanness is seen in terms of damage or harm to the body: eyes, teeth, and hair change to unnatural colors; unwanted body hair sprouts in private areas; and sheets are soiled with regressive bed wetting.

It is commonly observed that children sometimes explore delicate or taboo subjects symbolically "behind the smokescreen of playful rhyming" (Bronner 1988, 52). Whereas that is certainly true, on occasion their rhymes are remarkably blunt, as with this adaptation of "Colonel Bogey" that addresses its touchy theme with candor:

> Herman, look what you've done to me,
> Herman, I think it's pregnancy,
> Herman, you put your sperm in,
> And now it's Herman, and Sherman, and me. (Mahoney 1994, n.p.)

The self-referential monologue of a girl confronting the boy who has impregnated her, this rhyme depicts an unusual narrative stance. Lines 1–3 are directed at Herman alone, but we are drawn in as voyeuristic spectators to what would normally be a very private conversation. The speaker accuses her partner, speculates as to the consequences of their behavior, and reviews explicitly the activity that led them to that point (some versions offer a more decorous third line: "Herman, you were determined"). Taken together, the verses demonstrate considerable agency on the part of the speaker: she has confronted her partner, held him accountable for his actions, and proceeded on the expectation that they would naturally form a family unit with the newborn. That this version was an observed exchange between two ten-year-old girls is instructive, bearing out theoretical claims that children's play is an arena of practice and preparation for adulthood (e.g., Knapp and Knapp 1976; Bronner 1988, 32ff.; "Children's Playground Games" 2009–2011, 10). These girls are contemplating in song the connection between sex, pregnancy, and birth at a moment just before those things will have monumental importance in their lives. Alford could scarcely have imagined his patriotic march functioning as the underpinning for such lyrics nearly a century after its composition.

MILITARY REDUX

While the march has been widely adopted by civilians, including children, its martial applications were not exhausted by 1945. Some early baby boomers who had been exposed to WWII versions found themselves fighting in Vietnam. There, the song was reinvigorated with new themes and rhetoric suited to the context—like the following lyric that had more to do with the waning morale of American GIs than with demoralizing the enemy:

> Re-up, and buy a brand new car,
> Re-up, show what a fool you are,
> Re-up, I'd sooner throw up,
> I'd sooner throw up than re-up today.

When discharged, a soldier in the final stages of out-processing would see a recruitment counselor, whose unenviable job it was to encourage reenlistment. Personnel who extended for six or more months were entitled to thirty days' special leave to any place in the world, and as further enticement they were offered a reenlistment bonus, several thousand dollars exempt from taxation. According to the 25th Infantry Division reenlistment office, and reported in *Tropic Lightning News,* the division's weekly newsletter in Vietnam, the highest reenlistment bonus offered as of August

1968 was $9,540 ("SSG Jones Sets High In Re-up Bonus" 1968). As a point of comparison, the US Census Bureau reported that $8,600 was the median income of American families in 1968. It certainly would have been possible then to purchase a brand-new car with the reenlistment bonus. Nevertheless, it was an extremely difficult sale for recruiters to make. One blogger recalled the aphoristic reply that "short timers" would typically give when asked if they had thought about reenlistment: "Yes. Thought about it. Laughed about it. Forgot about it." The song's purpose, of course, was to steer GIs away from the redoubtable decision of reenlistment; in that sense, "Re-up" functioned as a sort of public-service announcement among fellow recruits. It is a tortuous path that brought "Colonel Bogey," an imperial military march with patriotic flair, to Vietnam—into the singing repertoire of American GIs who were counting the days until their discharge. This single melody became a sounding board for antithetical ideologies. Soldiers singing "Re-up" in the late 1960s may easily have had fathers who in WWII sang "Hitler Has Only Got One Ball" which, lewd though it was, elevated the morale and determination of Allied troops by ridiculing the common enemy. Conversely, "Re-up" is a picture of ambivalence: a jaunty military march conveys a message of antimilitaristic cynicism. Each version reflects the ethos of its time.

The utility of "Colonel Bogey" as a soldier's plaything is still in evidence. A video posted on YouTube in 2010 (with the heading "Explosions, Predator missiles, etc. in Iraq"), for example, shows footage of real-life explosions from the Iraq War. In what seems surely intended as ironic, the two-and-a-half-minute clip takes as its soundtrack the familiar *River Kwai* whistling arrangement of "Colonel Bogey." The bouncy and triumphant tune provides the backdrop for images of violent destruction, and soldiers' voices can be heard celebrating the blasts; war is depicted as glorious, victimless fun. The poster invokes the film clearly by tagging the video as the "*River Kwai* March." To be sure, explosives also figure prominently in *River Kwai*, with its unforgettable climactic scene of demolition. But there is playful manipulation at work in the video: seemingly incompatible sensibilities are merged when cheerful whistling accompanies conflagration. Here, as before, we see the multivalent social uses of the "Colonel Bogey March"—which in performance can be simultaneously both majestic and irreverent, patriotic and subversive.

The evocative melody has endured, but perhaps not exactly in the manner that Alford would have preferred. "Colonel Bogey" is arguably the most famous march ever composed—and that is due in no small part to the persistent folk-lyrical ruminations on the alleged testicular oddities of the Nazi high command. "To this day," argues John Trendell, "a mere reference to the opening notes of the first four bars [of "Colonel Bogey"] is sufficient to 'cock a snook' [a gesture of disdain] at any ill-favored or potential enemy" (1991, 98). The comically derisive lyrics may indeed have played some role in boosting the morale of Allied troops, but the popular appeal of "Hitler Has Only Got One Ball" has long outlasted that utilitarian function. The catchy little melodic motifs that made "Colonel Bogey" a great military march also created the perfect conduit for terse humorous lyrics, and so a host of disyllabic names and nouns—bollocks, bullshit, Hitler, Göring, Himmler, Goebbels, and later Comet, vomit, Herman, and Sherman—all found a natural poetic home. The lasting legacy of the "Colonel Bogey March" in many ways is indebted to these lyrics and their pop culture manifestations.

WORKS CITED

A Chat with Lottie Collins." 1895. *The Era*, 10 August.

Ainsztein, R. 1969. "Review of *The Death of Adolf Hitler: Unknown Documents from Soviet Archives*, by Lev Bezymenski." *International Affairs* 45 (2): 294–5. http://dx.doi.org/10.2307/2613016.

Bezymenski, Lev. 1968. *The Death of Adolph Hitler: Unknown Documents from Soviet Archives*. New York: Harcourt, Brace & World.

Bronner, Simon. 1988. *American Children's Folklore*. Little Rock, AR: August House.

Brophy, John, and Eric Partridge. 1965. *The Long Trail: What the British Soldier Sang and Said in the Great War of 1914–18*. London: Andre Deutsch.

Brownlow, Kevin. 1996. *David Lean: A Biography*. New York: St. Martin's.

Busby, Roy. 1976. *British Music Hall: An Illustrated Who's Who from 1850 to the Present Day*. London: Paul Elek.

Children's Playground Games and Songs in the New Media Age. 2009–2011. Centre for the Study of Children, Youth and Media; Arts & Humanities Research Council. Swindon, UK: AHRC.

Cleveland, Les. 1994a. *Dark Laughter: War and Song in Popular Culture*. Westport, CT: Praeger.

Cleveland, Les. 1994b. "Singing Warriors: Popular Songs in Wartime." *Journal of Popular Culture* 28 (3): 155–75. http://dx.doi.org/10.1111/j.0022-3840.1994.2803_155.x.

Davies, Christie. 2001. "Humour Is Not a Strategy in War." *Journal of European Studies* 31: 395–412.

Ellis, Bill. 1989. "When Is a Legend? An Essay in Legend Morphology." In *The Questing Beast: Perspectives in Contemporary Legend IV*, ed. Gillian Bennett and Paul Smith, 31–53. Sheffield, UK: Sheffield Academic Press.

Gervais, Ricky. 2007. *Ricky Gervais Live 3—Fame* [DVD]. Dir. Dominic Brigstocke. Universal Pictures Video.

Graves, Richard. 1999. "The Real Colonel Bogey." *Music and Vision* (April 7): n.p.

Knapp, Mary, and Herbert Knapp. 1976. *One Potato, Two Potato: The Folklore of American Children.* New York: Norton.

Macdonald, Lyn. 1993. *Somme.* Harmondsworth: Penguin.

Mahoney, Rosemary. 1994. "Scenes from Central Park." *Johns Hopkins Magazine* (September): n.p. http://www.jhu.edu/jhumag/994web/global1.html.

Margolick, David. 2005. *Beyond Glory: Joe Louis vs. Max Schmeling, and a World on the Brink.* New York: Knopf.

O'Brien, Donough. 2000. *Fringe Benefits.* London: Bene Factum Publishing.

Phillips, Gene D. 2006. *Beyond the Epic: The Life and Films of David Lean.* Lexington: University Press of Kentucky.

Redlich, Fritz, MD. 1999. *Hitler: Diagnosis of a Destructive Prophet.* New York: Oxford University Press.

Richards, Jeffrey. 2001. *Imperialism and Music: Britain 1876–1958.* Manchester: Manchester University Press.

Roberts, Andrew. 2008. "Did Hitler Really Only Have ONE Testicle? A Historian Sorts the Extraordinary Truth from the Far-Flung Myths about the Fuhrer." *The Mail Online* (London). http://www.dailymail.co.uk/news/article-1087380/Did-Hitler-really-ONE-testicle-A-historian-sorts-extraordinary-truth-far-flung-myths-Fuhrer.html (posted November 20).

Rosenbaum, Ron. 2008. "Everything You Need to Know About Hitler's 'Missing' Testicle: And Why We're So Obsessed with the Fuhrer's Sex Life." *Slate.com.* http://www.slate.com/articles/life/the_spectator/2008/11/everything_you_need_to_know_about_hitlers_missing_testicle.html (posted November 28).

Rosenbaum, Ron. 1998. *Explaining Hitler: The Search for the Origins of His Evil.* New York: Harper Collins.

Sherman, Josepha, and T.K.F. Weisskopf. 1995. *Greasy Grimy Gopher Guts: The Subversive Folklore of Childhood.* Little Rock, AR: August House.

Smith, Neville (writer), and Jane Morgan (producer). 2001. "Dear Dr. Goebbels" [radio broadcast]. BBC, November 30.

"SSG Jones Sets High in Re-up Bonus." 1968. *Tropic Lightning News* 3, no. 33 (August 12): 1.

Trendell, John. 1991. *Colonel Bogey to the Fore: A Biography of Kenneth J. Alford.* Dover: Blue Band Magazine; printed by A. R. Adams & Sons.

Waite, Robert G.L. 1977. *The Psychopathic God: Adolph Hitler.* New York: Basic Books.

11

Soldier Snaps

Jay Mechling

In March 2011, the German magazine and news organization *Der Spiegel* published a series of graphic photographs that had been seized by the US Army in their months-long investigation of five soldiers charged with murder and conspiracy in the deaths of three unarmed Afghans in 2010. *Rolling Stone* magazine later picked up on the story for its print and electronic editions (Boal 2011). The photos are gruesome in their depiction of dead bodies and body parts, but just as disturbing are the smiling faces of the American soldiers. The shocking amateur photos taken by American soldiers guarding prisoners at the infamous Abu Ghraib in 2004 depicted the humiliation of the prisoners by smiling guards (Sontag 2004), but the 2011 photos are even more shocking. Why did that "kill team" take those photographs and keep those digital images on their phones? What possible pleasure, if pleasure is the right word, would these soldiers experience looking again at those photographs and being reminded of the events?

I shall return to these questions below, where I analyze a number of vernacular images of soldiers "playing" with death in various ways. To begin this chapter on the meanings of vernacular photographs by soldiers, sailors, Marines, and fliers (hereafter simply called "soldiers"), I want to note first the significance of the digital technological revolution, such that almost every soldier has in his or her pocket a cellular phone that takes pictures and, in some cases, an actual digital camera. This technological development shifted the control of the visual narratives of the war from the government and professional photojournalists to the soldiers themselves. Those fighting the wars already had taken some control of the verbal narratives, through Internet

DOI: 10.7330/9780874219043.c11

blogs, often shut down by the military, and in letters home. But the power of images to represent wars, such as Joe Rosenthal's famous photograph of the flag raising on Iwo Jima and Nick Ut's photograph of a naked Vietnamese girl running from the napalm bombing of her village, means that in some ways the control of the soldiers over the visual images of the wars in Iraq and Afghanistan has more impact than the written accounts. Wisely or unwisely, we still attribute authenticity to the amateur snapshot, we still believe that the snapshot tells unfiltered truths about war experiences.

Photojournalists certainly have recorded everyday events in war theaters; some of the earliest photographs are of the Crimean War and of the American Civil War. By the Spanish-American War (1898), the professional photojournalist had become integral to the reportage of war (Moeller 1989), and through America's wars some of the finest documentary photographers have worked under fire, often at great risk. Some have died. The deaths of war photographers Tim Hetherington and Chris Hondos in the besieged rebel Libyan city of Misrata in April of 2011 reminded people of the risks these war photographers take (Chivers 2011). This history of American professional war photography is also a history of government censorship and even of "faked" photographs (Roeder 1993). Historians and others have documented and analyzed this history of professional war photography, and while that body of imagery is not the topic of this chapter, professional war photography does shadow our gaze at soldier snapshots. Most of us see war photographs by professionals before we see the amateur photographs, so the language of war photography—the iconic images, the conventions of subject matter and of framing, the nearly pornographic capturing of moments of human pain and death—is familiar to us and is something we bring to our viewing and understanding of the soldier snaps.

The vernacular photographs by the soldiers themselves had to await the invention by George Eastman in 1888 of the small, light, relatively inexpensive camera and the roll film carried in it. The term *snapshot* comes from hunting and no doubt was adopted in vernacular photography to reflect the same sense of a spontaneous, hurried shot in the field. That spontaneity is one of the features that lends the snapshot its aura of authenticity.

The military snapshots I have collected and seen at vintage photography shows and on Internet auctions begin with the Spanish-American War of 1898; by World War I, snapshots are common. One must assume that there are hundreds of millions of these vernacular photographs in existence; doubtless many more have been discarded. I am working with a very small sample of images, given the enormous domain of soldier snaps. Yet, I am convinced that we can see in even this small sample some of the patterns and conventions that characterize soldier snaps.

WHY LOOK AT SOLDIER SNAPS?

In her brilliant essay on war photography, *Regarding the Pain of Others* (2003; see also Sontag 2004), Susan Sontag reminds us that we bear a moral responsibility when we look at images of humans in pain. We have to earn the right to view those images, she says, by which she means (in part) that the images should impel us to act to alleviate the sources of the pain and suffering. Sontag writes mainly about well-known, iconic images, but her lesson applies to soldier snaps as well. Why look at these photographs?

Soldier snaps usually lack the drama of professional war photography. Professional war photographers often capture the heat of battle. Soldier snaps do so infrequently. Instead, like war memoirs and autobiographies, they testify to the truism that being in a war is 95 percent boredom and 5 percent sheer terror. The soldier's camera comes out during those boring times, but the truth telling of these images is as important as the action photos by professionals.

Perhaps I can bring some insight into the use of these images by saying what happens when I look at the soldier snaps. Surely some of the pleasure of viewing soldier snaps—and that is part of the Sontagian puzzle here, that I take pleasure in viewing these images—is that I am looking at ordinary people who find themselves in extraordinary circumstances. Having never experienced war myself, but believing the testimony that war creates unique existential moments and consciousness, I gaze at the people in these photographs, looking for some sign of that consciousness. They know something I don't know, but I know something they don't know. I know how the war came out, for example. And although I might not know who in the pictures survived the war and who did not, I conceivably could know.

I have read enough war memoirs, autobiographies, and oral histories to understand a few truths about war experiences by those who fought. One I've mentioned already, that war can be mostly boring and occasionally terrifying, and that the terrifying parts turn some people into "adrenaline junkies" who are drawn back into danger from safety and who sometimes cannot return easily to boring, everyday civilian life. Another is that soldiers fight not for ideas, not for God and country, not for motherhood and apple pie, but for the soldier to the right and the soldier to the left. The greatest fear reported in the war memoirs is the fear of letting down a comrade and of being responsible for their deaths and injuries. This is why "unit cohesion" comes up so often in the policy debates about women and gays in the military. Still another truth that dawns when you've read enough memoirs is that every soldier has his (or her) story, that the stories differ because experiences differ, and that the stories differ because of the frailty of human memory. And people lie. John Crawford entitled his Iraq War memoir *The Last True Story I'll Ever Tell* (2005) because he plays with the reader's tendency to take war stories as true, and in this regard his book resembles Tim O'Brien's famous Vietnam war memoir, *The Things They Carried* (1990), and a string of memoirs back to Company K (March 1933).

The war memoirs also tell this disturbing truth: war is irrational. Death is irrational. People live and die, emerge whole or physically and mentally wounded, not according to some ledger of justice and injustice but according to luck. Good luck, bad luck, dumb luck. You're talking to your friend one moment and his brains spatter your clothing and shoes the next. Everyone else in your Humvee is killed or wounded by an improvised explosive device (IED) on a road in Iraq, but you walk away without a scratch. That is the existential truth in the stories the war memoirists try to tell us, and that is the existential truth lurking behind the smiles on the faces in the soldier snaps.

As a male and as someone who writes about masculinity, I also take pleasure in recognizing the evidence of male friendship I see in many images. Some soldier snaps are by and about women in the military, but I am mainly interested here in the men in the photographs. As I show below with some specific snapshots, there is a familiar playfulness and intimacy of male friendship captured in the images, male bonding captured occasionally in fiction but powerfully in photography. Collecting images taken across a century shows some remarkable continuities in the play and body language of American male bonding.

As a folklorist I also take pleasure in seeing in the snapshots evidence of the folk cultures soldiers and sailors create in their small friendship groups. People use folklore to make their lives more meaningful and, sometimes, more bearable. One aspect of this is that folklore is more often expressive than it is instrumental; even the everydayness of folklore can have aesthetic dimensions. Folklorists know that people try to create beauty and meaning even in the most dire circumstances. So as a folklorist I'm interested in the military folk cultures not just as another case of male folk groups but as a confirmation that folklore really can comfort and bolster people as they face horrible circumstances and fates.

Photographs do not capture the everyday oral cultures (stories, jokes, proverbs, taunts, etc.) of these groups, though some snapshots capture the gatherings and settings (e.g., in barracks, in camp) where the participants perform their oral lore. Photographs best record customary and material culture. Visual images of customary lore bear a more complicated relationship to the behavior being documented. Is a photograph of soldiers playfighting, for example, merely an illustration of a custom that can be understood just as well though the verbal description, or does a photograph of that customary lore actually add something to our understanding of the custom? This is the difference between a snapshot as an illustration of something we know already through other evidence and a snapshot as an evidentiary "text" to be read or interpreted for new information and understandings, including some surprises not gleaned from other sorts of evidence. Some of the snapshots I examine here are useful as illustrations, but others interest me as texts, as evidence of some emotional truth all the words in the memoirs cannot tell.

My initial approach to making sense of the soldier snaps I have readily at hand was to make a taxonomy of sorts of the subject matter in the photographs. For example, many snapshots show soldiers with their "buddies." Others show the living spaces of the soldier, the bunks, the pinups on the wall, the personal artifacts that make "my space." In some snapshots soldiers are posing with native people, sometimes beside a topless native female Pacific Islander. Soldiers pose with pet animals—a tame monkey adopted by the group, perhaps an adopted stray dog or cat. Soldiers apparently thought it important to photograph the everyday work they did, from peeling potatoes to mopping floors to doing their laundry in large tubs. We see in the snapshots the soldiers at play, sometimes in card games, dice games, chess, and checkers, sometimes in games of baseball, football, or volleyball. We see soldiers wrestling and boxing.

The problem with organizing my discussion around these thematic categories is that some of the most interesting photographs combine these subjects and themes, so instead I shall approach my task of making sense of soldier snaps by focusing on a relatively few photographs that seem to me rich with information about the ordinary and extraordinary aspects of the folklore and folklife of the soldiers.

11.1. A "buddy hug."

I begin with a snapshot (figure 11.1) of two soldiers sitting on the ground and embracing in a "buddy hug."

There are many such snapshots, including some of two or even three buddies crammed into a photo booth. Ibson (2002) explores with great insight some of the possible meanings of these intimate poses between men in twentieth-century vernacular photography, as Deitcher (2001) does for photographs of men in the nineteenth century. Ibson notes that some of the men in the snapshots he studies are gay, but his more important point is that men of all sexual orientations derive pleasure from the embraces and other physical gestures of intimacy we see between men in the snapshots, including men in the military. Ibson writes at length about how to read such photographs as evidence of men's changing relationships in the twentieth century. Ibson notices that a great deal of heterosexual male touching was permitted in the photographs, both professional and vernacular, during WWII, but the poses change dramatically when the war ends. Men start posing with more distance between their bodies and without the touching, as if the war gave license to certain sorts of male intimacy that peacetime and the homophobia associated with the cold war did not permit.

As the graffiti on the railroad car in figure 11.2 shows, soldiers are fond of writing and drawing on things.

Soldiers often made humorous or more serious signs taunting the enemy. Soldiers and sailors paint totems on their weapons. Pilots customarily name their planes and paint "nose art" on the fuselage. These folk paintings name the plane, but they also serve as charms meant to protect the

11.2. The female body: a common motif in soldiers' graffiti.

fliers. Ships and submarines also have totemic paintings. Soldiers and sailors paint bombs, artillery shells, and torpedoes with "messages" to the enemy.

Snapshots such as figure 11.3 record the hazing and pranking common in all-male folk groups (Mechling 2009).

Soldiers play traditional pranks closely resembling those kids play in summer camps, often involving sleeping victims. Pranking and hazing flow into each other, from birthday spankings to more ritualized hazing of new members of the group. The Navy (including Navy and Marine Corps fliers) seems to have the most ritualized forms of hazing of all the services. Bronner (2006) has written the most thorough description and thoughtful analysis of the old naval tradition of hazing sailors on their first experience crossing the equator (figure 11.4), and I have little to add to his interpretation.

What is important to see—and this is where the snapshots prove invaluable—is the smiles on the faces of those doing the hazing and of those being pranked or hazed. Masculinity is such a fragile construction that it must be tested and proven on a daily basis (Mechling 2005). The paradox of play is that in the play frame (Bateson 1972 [1955]) words and behaviors do not mean what they would mean outside of the play frame. The seeming humiliation of pranking and hazing actually serves male bonding; hence the smiles.

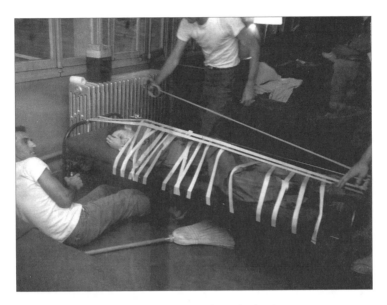

11.3. A sleeping Vietnam-era GI taped into his bunk.

Two categories of soldier snaps seem to me to be extremely revealing about the psychological state of the soldiers, and I shall devote the remainder of this chapter to a discussion of photographs revealing these states. Folklorists rely on the notion that folk traditions provide more impersonal, indirect, and symbolically coded ("displaced," in Freudian terms) resources for dealing with social tensions in the group and psychological tensions or anxieties in the individual (Abrahams 1968). Hazing serves to manage social tensions among soldiers, for example, while good luck charms in battle serve to allay psychological anxieties. So I propose to read some of the soldier snaps in order to understand the ways soldiers use their everyday folk traditions to handle the psychological stresses of war. The first issue concerns nudity and touching, the second "deep play."

MALE TOUCHING AND NUDITY

Scholars who study the creation, routine maintenance, and crisis maintenance of masculinity in American culture point to the many ways the male body condenses in one symbol a society's ideas about the nation. Metaphors using the body pervade our everyday language about other spheres, such as politics and society. The body serves as a perfect model for the bounded

11.4. A "crossing the equator" hazing ritual: a fake operation.

system, with inside and outside and taboos concerning substances (food, drink, feces, urine, etc.) that cross the dangerous boundaries (the liminal zones) between inside the body and outside (Douglas 1966). Building on this idea, historians and interpretive social scientists "read" the male body in the United States as a condensed symbol of the nation and of the qualities desired in the nation. Bordo (1999), for example, reads back-and-forth between the male body and the society, between the symbolic power of the phallus in a patriarchal society and the real, vulnerable penis.

This interpretive clue helps make sense of the many snapshots in which the male body seems to carry the symbolic burden for allaying the soldier's anxiety. Consider figure 11.5, a "pyramid" of WWII-era soldiers in camp.

Such pyramids and their variations show up quite often in the vernacular photography of men's friendship groups, including soldiers and Boy Scouts. A peculiarity of the human pyramid is that it is a folk custom apparently created specifically to be photographed. Doubtless gymnastics originated such body formations deep in the past, but boys and young men do not normally create such pyramids in everyday, natural settings. The pyramid

11.5. A pyramid of World War II–era soldiers.

and the snapshot, then, provide a folk formula that might "mask" the real motive of touching and being touched. Here is the buddy hug extended to the whole group. Moreover, I would count this pyramid tradition as an example of "rough-and-tumble" play because in most cases the pyramid comes apart by collapsing on itself into a tangle of male bodies.

Snapshots of rough-and-tumble play also add texture and insight into what we know from war memoirs and films. War memoirs are filled with accounts of traditional rough-and-tumble play. In his memoir of the First Gulf War, *Jarhead*, Anthony Swofford explains two of these customs in the middle of describing how the men of his unit began misbehaving when a reporter was trying to interview and film them for a story. They are playing a game of touch football, comically wearing their gas masks, when the play slips quickly into a game of tackle (Swofford 2003, 20). Swofford then goes on to describe the tradition of the "field fuck," "an act wherein marines violate one member of the unit, typically someone who has recently been a jerk or abused rank or acted antisocial, ignoring the unspoken contracts of brotherhood and camaraderie and esprit de corps and the combat family. The victim is held fast in the doggie position and his fellow marines take turns from behind" (20–21). The reader really doesn't need a snapshot to visualize the scene (much of this appears in the theatrical film based on the memoir), Swofford describes it so well. But a combination of the words and images of rough-and-tumble play among soldiers tells us something about the human psychological and social needs met by this traditional form of play.

It is worth making a few more analytical points here about rough-and-tumble play. First, there are deep biological and evolutionary reasons for rough-and-tumble play (playfighting) in mammals. Playfighting puzzles sociobiologists because it seems to have so many negative consequences for selective success. Besides the expenditure of valuable calories in play-fighting, the playing mammals may be injured and may be distracted from vigilance toward predators; yet, mammals playfight. In part playfighting can be explained as anticipatory socialization, as an occasion to practice and hone skills needed later in life. But some scholars (e.g., Sutton-Smith 1997; Burghardt 2005) argue that we should not overlook the pure plea-sure of play, that the evolution of the enjoyment of the consummatory act of play is every bit as important as other, more plainly instrumental evolutionary advantages.

A second important point about the functions and meanings of play-fighting is that it provides early instruction in the relationship between fan-tasy and reality. There is a reason why Gregory Bateson (1972 [1955]) uses the example of an animal playfight to show how the play frame works. "The playful nip denotes the bite," writes Bateson (180), getting at the funda-mental paradox of play, "but it does not denote what would be denoted by the bite." Thus the mere act of sharing the play (fantasy) frame signals to the players that in reality they have a close, trusting relationship.

Third, playfighting serves crucial psychological and social functions in male folk groups. American culture generally forbids heterosexual men's touching each other except in carefully framed ways. At the same time, humans (like other mammals) seem to need to be touched. American males solve this dilemma by engaging in rough-and-tumble play. Horan (1988) noted that the boys in the residential group home where he did his folklore fieldwork used playfighting to negotiate power, but also as an acceptable form of touching, especially for boys for whom touching was

11.6. World War II–era soldiers on a latrine.

highly problematic due to their experiences with sexual and physical abuse as children.

The frequency of male nudity in soldier snaps draws even more attention to the male body as a symbol and, potentially, as the site for expressing and quelling the soldier's psychological anxieties. One sort of seminude picture is of a soldier or group sitting on a latrine (figure 11.6).

In most cases the soldiers caught on the latrine are smiling, indicating they take no offense at being photographed in that vulnerable position. Sometimes the latrine pictures are intentionally humorous (figure 11.7).

A few things are going on here. First, personal privacy is scarce in military encampments; these snapshots both acknowledge that fact and may help defuse embarrassment by drawing attention to the public display of a bodily function usually performed in private, a common cultural strategy for dealing with anomalous elements (Babcock-Abrahams 1975). Second, Douglas's (1966) point about clean/dirty as a central, powerful binary that can be represented and negotiated with the human body as symbol draws our attention to the "dirt" here, namely, feces or, in the folk parlance, "shit." Memoirs and even documentary films about the Gulf War and the wars in Iraq and Afghanistan are filled with testimony about the smell of burning shit. "Shitter duty" involves pulling out the barrels of shit placed under latrine holes, pouring fuel oil into them, setting the mixture afire, and stirring the burning shit with a long stick. The memoirists and documentary filmmakers can never forget that smell.

So shit takes on great symbolic importance in the clean/dirty tension in the war theater. Lifton (1973) recounts several moments in group "rap sessions" with Vietnam vets indicating that dying a "clean death" was extremely important to these men. So the symbolic meanings of "shit" and of the dirt

11.7. World War II–era soldiers lined up for a latrine.

and mud they were constantly seeking to wash from their bodies were very important to these soldiers. To be "in the shit" was (and is) a common folk phrase for being in dangerous battle, and Lifton's recognition of the ways the clean/dirty binary entered the vets' stories and their dreams helps me to see more clearly the emotional impact of O'Brien's story in *The Things They Carried* of Kiowa's death in a field of mud and excrement. After that event, the narrator of the story just wants a bath: "Nothing else. A hot soapy bath" (1990, 150).

So bathing is the second setting in which the nude male body is displayed. It was not until WWII that public mores permitted the publication of photographs of Americans soldiers and sailors in various stages of undress, from bare chests to men clad only in shorts or boxer underwear to full nudity (though, of course, not frontal nudity in general-circulation publications). As scholars have noted (Ibson 2002), mainstream magazines during WWII featured full-page advertisements showing nude and semi-nude male bodies, the most famous of these a series of ads for Cannon towels. Many of the photographs taken by Edward Steichen's Naval Aviation Photographic Unit feature nude sailors and Marines showering, bathing (including "helmet showers"), or skinny-dipping in lakes, rivers, and oceans (Bachner 2004, 2007). In almost all cases only the bare buttocks show, though in a very few cases we see genitals. Fussell (1975) makes the point in a section of *The Great War and Modern Memory* (1975, 299–309) called "Soldiers Bathing" that in the literature he analyzes, the common scene of

11.8. A jungle shower, 1945.

soldiers' bathing together naked is treated as idyllic, romantic, and homo-erotic. "The quasi-erotic and the pathetic conjoin in these scenes," explains Fussell, "to emphasize the stark contrast between beautiful frail flesh and the alien metal that waits to violate it" (299).

The professional war photographs were for public consumption and were part of the propaganda effort in WWII, but the vernacular photographs by the soldiers themselves show the same preoccupation with the male body in war. There are plenty of partially clothed and unclothed male bodies in my sample of snapshots, and the prevalence of these images tells us something about masculinity in the military.

The soldiers in these snapshots are obsessed with cleanliness. The most obvious explanation for the prevalence of these images is that being in the field during wartime is dirty business, literally as well as figuratively. Men might crawl through mud and go for days or even weeks without bathing. So we see many snapshots of men posing in just towels or fully naked on their way to or from showers. We also see photos of men showering naked, with both butts and genitals showing, depending on the snapshot (figure 11.8).

Sometimes the men treat the towels and other objects like props in a strip show, carefully concealing their genitals and smiling for the camera. There are many snapshots of soldiers skinny-dipping.

There are a few important things to say about the images of this nudity. First, the folklorist recognizes that a function of nudity in male groups is to affirm the assumption that those men are heterosexual. The meta-message of this stylized nudity is "We can be nude in each other's presence because

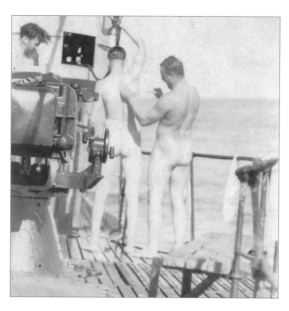

11.9. Submariners bathing one another.

we are heterosexuals and do not see each other's bodies as sexual objects."
The careful framing of male nudity and even tender male touching as heterosexual makes possible such snapshots as figure 11.9, of one nude sailor
washing the back of another.

Of course, even though the carefully framed, everyday male nudity
captured by the snapshots functions to construct all of this touching and
mutual bathing as heterosexual, we also can appreciate some of the alternative meanings of the homoerotic scenes. As Ibson (2002) and Bronner
(2006) say, military rituals like Crossing the Line are filled with homoerotic
symbolism, and the play frame of the ritual can "mask" (Sutton-Smith and
Kelly-Byrne 1984) other motives. If, as psychoanalytic gender theory suggests (Mechling 2001, 2005), the social construction of male heterosexuality requires the repression of the feminine side of the male's natural bisexual
self, then the homoerotic displays in a play or ritual time and space must be
providing the players some erotic pleasure.

Beyond this function of nudity in the male folk group, though, there
lies the question of the seemingly public ease with images of the nude male
buttocks, but not the genitals. The snapshots were for viewing by private
groups. Similar photographs in *Life* magazine and in magazine advertisements during WWII, though, raise larger cultural questions, especially since

this ease with the nude male body ends by the late 1940s (as Ibson 2002 notes) and reappears only in the fashion photography of Bruce Weber and others in the 1980s.

When I was writing an article on the meanings of paddling in fraternity hazing, I looked in vain for a scholarly treatment of the male buttocks as a symbol in American culture (Mechling 2008b). I cannot write that history in this short space, but I can suggest a few key points. We know from cultural histories of the male body, especially in American visual popular culture (excluding pornography), that the penis is covered and protected through most of the twentieth century (Weinberg 2004, 2008b). As I indicated, the male butt appears in images up until the early 1950s and then disappears, only to reappear in the 1980s. That hiatus has everything to do with the homophobia of cold war America, but the larger point is that the male butt stands for male strength and male strength stands for the nation. The muscled male thighs and buttocks are far superior signs of national strength than is the relaxed penis. The Nazis exploited the German cult of the male body to equate the muscled male body with the nation (Mosse 1996), and the Americans employed the same strategy. Soldiers' nude and seminude bodies came to represent the strength of the nation. Given these meanings, photographs showing muscled, military male bodies contribute to the message of the images: that America is strong. The snapshots show us that even the amateur photographers understood this.

There is more floating anxiety in these photographs here than the loss of privacy. In *On My Honor* (2001) I speculated (using Freud's Wolf Man analysis) that the prevalence of anality in the folk speech at an all-male camp might have to do with the repression of the feminine the boys and men accomplished toward the construction of the heterosexual male in that setting. In Freud's famous case and in subsequent psychoanalytic writing by others about anal fixation, the psychic "cost" of the repression of the feminine in the male reveals itself in male group folklore, including the jokes, pranks, hazing, and so on with a focus on the anus and feces. The prevalence of the latrine and the male butt in soldier snaps provides evidence, I think, of a pervasive anality in the folklore of the male combat group.

One more sort of display of the male body in the soldier snaps remains to be seen and analyzed. There are many photographs of men dressed as women, the female sometimes suggested only minimally, perhaps by a mop head as female hair or by an improvised bra (figure 11.10).

11.10. Cross-dressing with an improvised bra.

Sometimes the cross-dressing is more fully realized. Folk cross-dressing is rarely mentioned in printed sources or pictured in the professional photography, so this is another area where the snapshots provide unique evidence of the folklore. The Crossing the Line ritual documented and analyzed by Bronner (2006) always involves sailors in full female costume. Anyone who has seen *South Pacific* on stage or screen knows that soldiers and sailors commonly cross-dress for silly theatricals. This cross-dressing play is common in all-male groups, and I saw this sort of play in the Boy Scout troop I studied for my 2001 book, *On My Honor*.

The cross-dressing in all-male groups is a complicated matter (see Ibson 2002, 73–74). One understands why men must play female roles in the theatricals and skits in an all-male group, but the snapshots reveal lots of cross-dressing apart from folk theatrical performances. The cross-dressing is playful, meant to be funny and fun, and in the play frame the symbolic inversion creates the paradox of play—the female who is not a female, the male who is not a male. The play frame for the gender inversion paradoxically reinforces the masculinity of the player; presumably, in this context, only a man comfortable with his masculinity and firm in his heterosexuality would dress like a woman for the play space and time. The more masculine the man in everyday life, the more funny is the cross-dressing. The more feminine the man in everyday life, the more potentially edgy is the cross-dressing.

The behavior of the other men in the cross-dressing snapshots interests us some, as they often are laughing and groping the fake breasts and crotches of the men dressed as women. Alcohol may be involved in some of these snapshots. Finally, one wonders what effect the increasing presence of real women in the military and in field operations has on the cross-dressing play.

DEEP PLAY

Sutton-Smith and Kelly-Byrne (1984) adopt Geertz's (1973) notion of "deep play," which Geertz borrowed from Jeremy Bentham, to show how play can mask the irrational, dark drive for danger. The economist and philosopher Bentham was trying to understand irrational behavior, particularly play in which the stakes are so high that people should not be engaging in the play, but do nonetheless (Geertz, 1973, 432–33). Geertz sees "deep play" in the Balinese cockfight he so famously analyzes, and Sutton-Smith and Kelly-Byrne offer some examples of dangerous play they collected from their students.

Consider the snapshot "Kill the Beast" (figure 11.11). I did not title this photo; someone wrote "kill the beast" in ballpoint pen on its upper border.

In the snapshot, which appears to be from the Vietnam War era, a black soldier is holding a revolver and pointing it at the head of a white soldier, who is kneeling in front of the revolver with hands folded in a traditional gesture of prayer or begging. My first take on this photograph is that it falls into the "dark play" category of gun play (the category also encompasses knives and bayonets) I have found in so many soldier snaps from WWI to the present. But while many of those snapshots show signs of play—smiles, mock horror, and so on—the "Kill the Beast" snapshot might be signaling something more than simple playfighting. In his book, *Home from the War* (1973), his account of working with "rap groups" of Vietnam battle vets in the late 1960s and early 1970s, Lifton (a gifted psychoanalyst and psychohistorian) noted that the vets often used the word *beast* to describe what they had become as a result of what they did and witnessed in Vietnam. They carried a "beast within," a beast increasingly out of touch with the person's humanity, out of touch with other humans, and capable of sudden, irrational violence (see also Mogelson 2011).

11.11. "Kill the Beast."

So now the "Kill the Beast" snapshot invites another look and some interpretations beyond the recognition that this is dark play. Who wrote the caption on the border? Was this snapshot owned by the black soldier or the white one? Is the white soldier begging for his life or begging to be killed, put out of his psychological misery? And should we make anything of the racial difference? Is the snapshot one of symbolic revenge by people of color (African American, Asian, and Latino soldiers as well as the Vietnamese themselves) against the white man who uses people of color to kill other people of color? This is all still in the play frame, I imagine; I feel certain that the black soldier did not shoot the white one. But what did this photograph mean to them, especially to the soldier who kept the photograph until it entered the orphan snapshot market of collectibles dealers?

I selected the title of this chapter, "Soldier Snaps," quite early in my work on this project and, of course, I meant soldier snapshots. But on more than one occasion friends have told me that when I first revealed the title of the chapter, their thoughts immediately moved to "snaps" as the description of a soldier who has a sudden, violent mental breakdown. Reading the literature on post-traumatic stress disorder (PTSD) for another project

11.12. Dark play in World War I.   11.13. Dark play in World War II.

and hearing my friends' initial understanding of "snaps" led me back to the snapshots for a fresh look at their meanings and functions.

Folk traditions can provide a person with resources for dealing with the grief and anger that accompanies trauma and loss, and certainly soldiers feel that anger and grief when a comrade dies or is severely wounded. Soldiers in war zones see horrific things and sometimes must do horrific things; they return to civilian life quite changed. PTSD did not become an officially recognized psychological disorder (or injury, as Shay [2002] prefers to call it) until 1980, but American veterans have long suffered the symptoms called "nostalgia" in the Civil War, "shell shock" in WWI, and "battle fatigue" in WWII and the Korean War. By the Vietnam War, psychologists and psychiatrists were seeing and defining what came to be known as PTSD. Alongside these professional treatments, however, veterans have long treated themselves, often with alcohol or other drugs, sometimes with suicide. The psychological injuries sustained in war often lead the veteran to engage in

11.14. An officer's tent, with a skull on a stake.

unusually risky behavior (Welch 2009). Put differently, the veterans engage in "deep play."

In most cases the symptoms of PTSD do not wait for the return home. Shay (1994, 2002), Lifton (1973), and others who write about PTSD note the general lack of rituals of grief and trauma on the battlefield, including the policy of evacuating the bodies of dead soldiers as quickly as possible for the sake of the morale of the living. Lifton and Shay both believe that the lack of such rituals of loss lead to the "berserk state" (Shay 1994, chapter 5, "Berserk," 77–99) observed in veterans from classical Greece (e.g., Achilles in Homer's *The Iliad* and Ajax in Sophocles's play *Ajax*) to the present.

The literature on PTSD led me back to the soldier "snaps" (that word now laden with the double meaning) and, especially, to the photos that show soldiers engaged in "deep play." Consider, for example, figures 11.12, 11.13, and 11.14. These photos shock the viewer. The WWI-era soldier (figure 11.12) pointing his pistol at the head of an artillery shell (placed between his legs like a giant penis, no less) is engaged in dark play, certainly.

And then there are the photos (figures 11.13 and 11.14) with human skulls (presumably enemy skulls) as part of the tableaux, suggesting gallows humor, dark play, "deep play."

I now think that taking, keeping, and viewing these snapshots amount to a sort of folk therapy practiced by the soldiers on themselves in the war theater. As in all gallows humor, what is feared the most is put in the center of the imagery. Just as human skulls have appeared in Western art to remind us of our mortality and of the relative unimportance of this life, so the skulls in these photographs remind the living soldier of the thin line separating the living from the dead. The photographs prove at least a temporary victory over death, a temporary bit of luck against the larger irrationality of war.

Even the photographs of soldiers playing games in their downtime suggest a sort of self-therapy for the stresses and traumas of battle, though admittedly much tamer than the deep play portrayed by guns, knives, and skulls. Recall that Geertz adopted Bentham's notion of "deep play" to help him make sense of the Balinese cockfight, particularly of the "irrational" gambling that accompanies the cockfights. This puts into new light the games of cards, dice, checkers, and chess recorded in so many soldier snaps. Some of these games combine strategy and skill with a large element of chance, such as card games and dice games (e.g., craps). As we have noted, war memoirs testify to the disturbing role of chance (good luck and bad luck) in battle, how it is that military skills and strategies can be undone in a moment by chance. Games bring order to disorder—in a war theater games may be the only islands of order in a very disorderly world. That explains why games of strategy and skill (e.g., chess, most ball games) would be attractive; the players exercise control in the game world, control they may not feel on the battlefield. The games involving chance (cards, dice) have a different attraction, though; they test fate. They may provide in the game time and space a safe place to tempt chance, to see if skill and strategy really can work around chance. There is much more to be said about this, but I think the soldiers take snapshots of these leisure time games not merely to say, "Here's how we play when we're not fighting or working." The photographers understand at some level—probably unconscious—that playing these games is a very important, safe frame for exploring the relationships among skill, strategy, and chance.

CONCLUSION

My goal here has been not just to create a taxonomy of the vernacular photographs of soldiers, Marines, sailors, and airmen but to distinguish between those soldier snaps that merely illustrate some military folklore we already know about through other evidence and those snapshots that provide some unique evidence about military folklore and folklife. In two areas—deep play and bodylore—the snapshots provide unique evidence. What's more, the images do more than document the more hidden and masked aspects of military folklife. The snapshots make us realize how the very act of taking the photograph amounts to a folk custom that acts like other folk customs in addressing psychological and social anxieties. The snapshots bring some order to an irrational world of death, pain, and loss.

Returning to the examples with which I began—the Abu Ghraib photographs from Iraq and the "kill team" photos from Afghanistan—this realization that the taking of the pictures may be a folk custom meant to tame anxieties suggests that the soldiers in those photos knew that what they were doing to other humans was wrong. Sociopaths sometimes make it into the military, as might have been true in the "kill team" case (Mogelson 2011). But even in that case most of the participants are not sociopaths. I suspect that they took and saved the damning photographs not because they are sadists who enjoy the pain of others, but perhaps because the act of photographing the abuses and humiliations was meant to alleviate their anxieties and feelings of guilt. As Sontag and others writing about photography note, the camera sometimes puts a safe distance between the photographer and the scene, absolving the photographer from any responsibility for the human suffering being photographed. At the same time, these snapshots may give the soldiers a momentary look back at their beast within, creating a tension around the psychic trauma of the event.

Swofford (2003, 165) reports that the motto of many Marines in his platoon during the First Gulf War was "Fuck the suck," a catchy phrase that captured the fatigue, discomfort, and frustration felt by Swofford and his colleagues as they put up with the Corps, the desert, and the war. "Fuck the suck" inverts the common (since Vietnam) Marine Corps motto, "Embrace the suck," a proverbial phrase that advises the Marine facing a bad situation to welcome it and press on. I present as my last illustration figure 11.15, the sort of image the professional war photographer is unlikely to reproduce.

11.15. Fuck the suck?

For the folklorist, though, the picture condenses so much of the meaning that these snapshots add to our understanding of the experience of war by those who fought and fight. The body language of the soldier, his sitting in something like a fetal position, tells you that he has had enough. He wants to be left alone. At the same time, he raises not one but two middle fingers, a traditional obscene gesture. Is he saying, "Fuck you" to the photographer? Is he saying, "Fuck you" to the army and the war? Maybe he's just saying, "Fuck the suck."

WORKS CITED

Abrahams, Roger D. 1968. "A Rhetoric of Everyday Life: Traditional Conversational Genres." *Southern Folklore Quarterly* 32: 44–59.

Babcock-Abrahams, Barbara. 1975. "Why Frogs Are Good to Think and Dirt Is Good to Reflect On." *Soundings* 58: 167–81.

Bachner, Evan. 2004. *At Ease: Navy Men of World War II*. New York: Harry N. Abrams.

Bachner, Evan. 2007. *Men of WWII: Fighting Men at Ease*. New York: Harry N. Abrams.

Bateson, Gregory. 1972 [1955]. "A Theory of Play and Fantasy." In *Steps to an Ecology of Mind*, 177–93. Chicago: University of Chicago Press.

Boal, Mark. 2011. "The Kill Team: How US Soldiers in Afghanistan Murdered Innocent Civilians." *Rolling Stone*, March 27. http://www.rollingstone.com/politics/news/the-kill-team-20110327; http://www.rollingstone.com/politics/photos/the-kill-team-photos-20110327/0602176.

Bordo, Susan. 1999. *The Male Body: A New Look at Men in Public and Private*. New York: Farrar, Straus and Giroux.

Bronner, Simon J. 2006. *Crossing the Line: Violence, Play, and Drama in Naval Equator Traditions*. Amsterdam: Amsterdam University Press. http://dx.doi.org/10.5117/9789053569146.

Burghardt, Gordon M. 2005. *The Genesis of Animal Play*. Cambridge, MA: MIT Press.

Chivers, C. J. 2011. "'Restrepo' Director and a Photographer Are Killed in Libya." *New York Times*, April 20. www.nytimes.com/2011/04/21/world/africa/21photographers. html?_r=1&pagewanted=all.

Crawford, John. 2005. *The Last True Story I'll Ever Tell*. New York: Riverhead Books.

Deitcher, David. 2001. *Dear Friends: American Photographs of Men Together, 1840–1918*. New York: Harry N. Abrams.

Douglas, Mary. 1966. *Purity and Danger*. London: Routledge & Kegan Paul. http://dx.doi. org/10.4324/9780203361832.

Fussell, Paul. 1975. *The Great War and Modern Memory*. Oxford: Oxford University Press.

Geertz, Clifford. 1973. "Deep Play: Notes on the Balinese Cockfight." In *The Interpretation of Cultures*, 412–53. New York: Basic Books.

Horan, Robert. 1988. "The Semiotics of Play Fighting at a Residential Treatment Center." In *Adolescent Psychiatry: Developmental and Clinical Studies*, vol. 15, ed. Sherman C. Feinstein, 367–381. Chicago: University of Chicago Press.

Ibson, John. 2002. *Picturing Men: A Century of Male Relationships in Everyday American Photography*. Washington, DC: Smithsonian Institution Press.

Lifton, Robert Jay. 1973. *Home from the War: Vietnam Veterans, Neither Victims nor Executioners*. New York: Basic Books.

March, William. 1933. *Company K*. Tuscaloosa: University of Alabama Press.

Mechling, Jay. 2001. *On My Honor: Boy Scouts and the Making of American Youth*. Chicago: University of Chicago Press.

Mechling, Jay. 2005. "The Folklore of Mother-Raised Boys and Men." In *Manly Traditions*, ed. Simon J. Bronner, 211–27. Bloomington: Indiana University Press.

Mechling, Jay. 2008a. "The Cultural History of the Penis." In *The Cultural Encyclopedia of the Body*, vol. 2, ed. Victoria Pitts-Taylor, 384–90. Westport, CT: Greenwood Publishing Group.

Mechling, Jay. 2008b. "Paddling and the Repression of the Feminine in Male Hazing." *Thymos: Journal of Boy Studies* 2 (1): 60–75. http://dx.doi.org/10.3149/thy.0201.60.

Mechling, Jay. 2009. "Is Hazing Play?" In *Transactions at Play*, ed. Cindy Dell Clark, 45–61. Lanham, MD: University Press of America.

Moeller, Susan D. 1989. *Shooting War: Photography and the American Combat Experience*. New York: Basic Books.

Mogelson, Luke. 2011. "A Beast in the Heart of Every Fighting Man." *New York Times Magazine*, May 1, 34–41, 54, 62–63.

Mosse, George L. 1996. *The Image of Man: The Creation of Modern Masculinity*. New York: Oxford University Press.

O'Brien, Tim. 1990. *The Things They Carried*. New York: Broadway Books.

Roeder, G. H., Jr. 1993. *The Censored War: American Visual Experience during World War Two*. New Haven, CT: Yale University Press.

Shay, Jonathan. 1994. *Achilles in Vietnam: Combat Trauma and the Undoing of Character*. New York: Scribner.

Shay, Jonathan. 2002. *Odysseus in America: Combat Trauma and the Trials of Homecoming*. New York: Scribner.

Sontag, Susan. 2003. *Regarding the Pain of Others*. New York: Farrar, Straus and Giroux.

Sontag, Susan. 2004. "Regarding the Torture of Others." *New York Times Magazine*, May 23.

Sutton-Smith, Brian. 1997. *The Ambiguity of Play*. Cambridge, MA: Harvard University Press.

Sutton-Smith, Brian, and Diana Kelly-Byrne. 1984. "Conclusion: The Masks of Play." In *The Masks of Play*, ed. Brian Sutton-Smith and Diana Kelly-Byrne, 184–97. New York: Leisure Press.

Swofford, Anthony. 2003. *Jarhead: A Marine's Chronicle of the Gulf War and Other Battles*. New York: Scribner.

Weinberg, Jonathan. 2004. *Male Desire: The Homoerotic in American Art*. New York: Harry N. Abrams.

Welch, William M. 2009. "As Deaths Spike, Motorcycle Training Pushed for Troops." *USA Today*, May 6. http://www.usatoday.com/news/military/2009–05-06-soldierriders_N. htm#.

12

"America's Best"
Cultural Poaching on "Ballad of the Green Berets"

Tad Tuleja

Greatest military hymn of all times. your not AMERICAN *if this doesnt choke you up.*

—YouTube comment on "Ballad of the Green Berets"

What is distributed is not completed, finished goods, but the resources of everyday life, the raw material from which popular culture constitutes itself.

—John Fiske

In 1965, more than 180,000 American troops were stationed in Vietnam. That spring, responding to a Vietcong attack on the US army base at Pleiku, President Johnson approved the continuous bombing of North Vietnam known as Operation Rolling Thunder. Soon afterward, teach-ins spread across college campuses; protesters surged to rallies; journalist David Halberstam published a book with the unnerving title *The Making of a Quagmire*; and telegrams flooded the White House, those opposing escalation outnumbering by six to one those favoring it (Prados 2009, 112–32). Even staunch supporters of administration policy had to admit that, in that first year of serious US commitment, the American people were not on board with the war.

And the peaceniks had all the good songs. With the folk music revival in full swing, even AM radio, saturated by pop, soul, and surfer music, was airing antiwar songs like Pete Seeger's "Where Have All the Flowers Gone?" Bob Dylan's "Masters of War," Phil Ochs's "I Ain't Marching Anymore,"

DOI: 10.7330/9780874219043.c12

Buffy Sainte-Marie's "Universal Soldier," and Barry McGuire's "Eve of Destruction." With the exception of the McGuire song, these were not Top 40 hits, but they outplayed handily the few pro-military songs that even made it to the airwaves. If measured by Cash Box or Billboard ratings, Americans opposed to the Vietnam War far outnumbered those with pro-military sentiments. In 1965, hawks looked in vain for a patriotic hit.

The following year, they found it. The single most popular song of 1966—it spent five weeks at Billboard's number one position—was a poignantly pro-military anthem, "The Ballad of the Green Berets," which praised the courage of the US Army's Special Forces. It was cowritten by journalist Robin Moore, author of the best-selling book *The Green Berets*; and Special Forces staff sergeant Barry Sadler, who had been wounded in Vietnam and whose performance of the song on radio and television made him a celebrity. The song became the title track of Sadler's album *Ballads of the Green Berets,* which featured such other paeans to heroic sacrifice as "Letter from Vietnam," "The Soldier Comes Home," and "I'm a Lucky One." Adopted by John Wayne as the theme of his 1968 movie *The Green Berets*, it soon became a political touchstone, as unreflectively revered by hawkish patriots as it was unreflectively ridiculed by their detractors.

Because the song became so invested with the passions of the Vietnam era, it may be tempting to dismiss it as a period piece—a relic of the 1960s whose elegiac sentimentality can best be understood as a reflection of that turbulent time. But the song's afterlife belies this reading. In the four decades and more since its debut, "The Ballad of the Green Berets" has demonstrated a remarkable durability. YouTube contains dozens of video versions, and posted comments on them show that, decades after Sadler's death, the "Ballad" continues to inspire both respect and disdain.

In his autobiography, *I'm a Lucky One*, Barry Sadler proudly quoted a newspaper editor who believed that the "Ballad" would "live as long as 'Long, Long Trail,' 'White Christmas,' or any that you can name" (Sadler 1967, 180–81). We won't be able to judge the validity of that prediction for some time. But I suspect that as an emotional trigger, not just a historical footnote, Sadler's song will continue to be played and sung and wept over long after the events of Vietnam have been forgotten. In this chapter, I will try to defend this seemingly rash claim by exploring the song's long and still shimmering shadow.

THE BALLAD ITSELF: CORE VIRTUES

Let me look first at the song itself, particularly at one key to its popularity: its lean, poignant celebration of core military virtues. The first stanza introduces these virtues:

Fighting soldiers from the sky
Fearless men who jump and die
Men who mean just what they say
The brave men of the Green Beret

The dominant virtue here is the conventional one of courage. Green Berets, as exemplary fighters, are "fearless" and "brave." But they are also straight shooters in the idiomatic sense: men whose word can be relied upon because—unlike, say, dissembling politicians—they "mean just what they say." And they are unafraid of death. In fact, given the song's second line, it might almost be said that their role is to face death unafraid: they are, by definition, men who first jump (Green Berets are paratroopers) and then die.

A captious YouTube poster once pointed out that the Green Beret's actual role has never been to die but to prevail against the enemy. This echoes General George Patton's famous quip about the soldier's duty being not to die for his country, but to "make the other poor bastard die for his country." Sadler understood this, yet he chose to paint his comrades not as liberators or honorable men or victors, but as casualties—or, to be more precise, sacrificial victims. Like Roman gladiators, they are *morituri*—those about to die. Their willingness to do so is their signature virtue.

The song does celebrate other qualities. The chorus, for example, focuses on their elite professional status:

Silver wings upon their chest
These are men, America's best
One hundred men will test today
But only three win the Green Beret

And the middle stanza, picking up the courage theme, references the Green Berets' skills as "natural" backwoodsmen—a motif that has informed America's frontier mythology from the days of Natty Bumppo to those of John Rambo.

Trained to live off nature's land
Trained in combat, hand to hand

Men who fight by night and day
Courage take from the Green Beret

But elite status and backwoods savvy are ancillary qualities. The dominant virtue here—the virtue that provides the song's emotional core—is self-sacrifice, and more specifically, the willingness to die "for those oppressed."(The Special Forces' official slogan is De Oppresso Liber, rather freely translated as "To Free the Oppressed.")

Self-sacrifice is a cardinal heroic virtue. It is also the virtue that most easily captures the emotional loyalty of an audience—and the one most difficult for peacemongers to discredit. It is therefore (I do not mean this disrespectfully) a preeminently marketable quality. Sadler's co-lyricist, Robin Moore, was cagily aware of this fact. It was he who, in editing an early draft of the song, suggested its final stanzas (Sadler 1967, 174), in which dying "for those oppressed" becomes not only the Green Beret's "fate" but also a legacy passed on to the dead hero's son.

Back at home a young wife waits
Her Green Beret has met his fate
He has died for those oppressed
Leaving her his last request

Put silver wings on my son's chest
Make him one of America's best
He'll be a man they'll test one day
Have him win the Green Beret

The emphasis on sacrifice strikes me as curious in two ways. First, it reduces the complexity of Special Forces activity to a narrow cinematic band that is as unreal as it is sentimentally appealing. In the universe of Sadler's actual comrades, men function not only as elite warriors but—to give a few examples—as communications specialists, teachers, translators, medics, midwives, and engineers. Counterinsurgency is admittedly an important element of Special Forces operations, but the organization's primary business has never been to visit Rambo-like mayhem on the bad guys. In the words of one veteran Green Beret: "SF people would be among the last people to fix bayonets and attack a hill! They would be the first to say: there must be a better way! . . . They look at a whole situation and say: is what we're doing here smart? Is there a better way to win the hearts and minds than with bullets?" (quoted in Halberstadt 1988, 128). None of the "hearts and minds" focus is evident, of course, in the *Rambo* films. And none of it is evident in Sadler's musical tribute. In both of these venues, the Forces' creativity and political astuteness are effaced by stereotype.

Second, the emphasis on sacrifice sets the "Ballad" apart, emotionally speaking, from other songs celebrating military cadres. Since it has wide appeal among real Green Berets, we may consider it an unofficial service song, suitable for comparison with the military academy anthems. Those anthems pulse with optimism and activity. The Army's "Caisson Song" celebrates the invigorating sweat of an artillery march. "Anchors Aweigh," originally an Annapolis fight song, expresses the joy of beating Army on the gridiron. The "Air Force Song," despite a brief mention of possibly going down in flames, exults in the exhilaration of armed flight. The "Marine Hymn" expresses pride in the branch's history. In all of these examples, unit pride is celebrated triumphantly.

In Sadler's "Ballad," on the other hand, victory is invisible. If the song's dominant virtue is self-sacrifice, the audience's invited reaction to that virtue is not triumph or pride but somber acceptance. Even as he dies, the Green Beret enjoins his wife to outfit his child with silver wings, so that the sacrificial tradition may be passed on. This is an oddly sobering twist on service-song convention, yet one that Sadler heartily endorsed. In *I'm a Lucky One*, describing the song's popularity, he approvingly quotes an editorial from a North Carolina newspaper: "'Ballad of the Green Berets' causes chills to run up and down your spine. It has a mournful note that makes you see and feel war in all its strangeness" (Sadler 1967, 181).

In 1966, of course, that "mournful note" was devastatingly appropriate. Whether you were pro-war or antiwar, you could hardly fail to be moved by the growing rate of sacrifice, as eighteen-year-olds expired on the late-night news, the term "body bag" entered the American vernacular, and GI casualties continued to mount. By the end of that year, Special Forces units alone had lost 143 of their number in Vietnam. Sadler listed these at the end of his book (1967, 187–91). In this atmosphere, patriotic sympathy for "those about to die" was not an entirely unexpected response.

A NEW PATRIOTIC ANTHEM

That response sent ripples through a music industry that, until that point, had not been conspicuous in its support of the troops. Sadler's recording came out in January; within months, it was covered by several big-name performers. Among the first was Kate Smith, whose signature song was "God Bless America" and who had been the doyenne of chest-thumping patriots since the 1940s. Teresa Brewer included the "Ballad" on her album *Songs for Our Fighting Men.* Ken Darby provided a crooner's version for the John Wayne movie, Johnny Paycheck a country version in a Carnegie Hall live album. The king of the twangy guitar, Duane Eddy, did an instrumental version on his album *The Biggest Twang of Them All.* The song also appeared in lounge versions, collections of "Sousa" marches, and in various compilations of patriotic tunes.

I want to make two points about this flurry of covers. First, it seems clear that, at a moment when opponents of the war were struggling with an old pacifist conundrum—how to condemn a war without dishonoring warriors—the Sadler/Moore ballad effectively appropriated ownership of sympathy for the troops, elevated their sacrifice into a political virtue, and thereby defined opposition to the war as not only unpatriotic but potentially treasonous—as, in fact, an attack on the troops themselves. This type of propagandistic blurring is standard practice in wartime, and it is frequently effective. There is no more potent emotional argument for continuing a fight than to claim that stopping it will invalidate previous sacrifices. ("Our boys will have died for nothing.") In making that argument, "Ballad of the Green Berets" set the stage, I suggest, both for a more vigorous defense of the war and for such later bloody-flag favorites as Merle Haggard's "Okie from Muskogee" and "The Fightin' Side of Me."

The second point I want to make concerns the transformation of the song from a topical hit into a patriotic standard. This happened quickly, and the song is still understood as a fit companion not merely to Vietnam-era songs but to flag-waving standards like "America the Beautiful," "The Star-Spangled Banner," and the service marching songs. Ace Collins's *Songs Sung Red, White, and Blue* (2003), for example, tells the "stories behind America's best-loved patriotic songs." Its alphabetical arrangement has "Ballad of the Green Berets" tucked in between "The Army Song" and "Battle Cry of Freedom," in a book that also includes the other service songs, "My Country 'Tis of Thee," "Yankee Doodle," and "This Land Is Your Land." An odd mix from an ideological perspective; not so odd from that of conventional sentiment.

Patriotic standards tend to evoke the lump-in-the-throat response from an iconic rather than indexical frame of reference. The teenagers who sing "This Is My Country" in high school auditoriums are moved by its bland triumphalism, not its utility as World War II propaganda; Americans who could not find the shores of Tripoli on a map can still be stirred emotionally by "The Marine Hymn." This is why, as the trauma of Vietnam recedes from public memory, the "Ballad" has acquired a veneer of nonspecific reverentiality. It has become not merely a pained memento of a troubled era but one more jewel in the diadem of national greatness—a reliable tearjerker with the power to evoke not historical recall but a collective sense of "Americanness" that is unexamined, perhaps unexaminable, and cut loose from history.

POACHING AND PRODUCERLY BEHAVIOR

But if the emotional core of the "Ballad" touched many Americans, others had quite different responses, and these responses generated musical interpretations that ranged from simple adaptation to outright parody. Rather than consuming the song passively, many engaged in what John Fiske, adapting Michel de Certeau's notion of "poaching," calls "producerly" behavior. A producerly consumer, rather than absorbing unreflectively the texts a dominant culture industry offers, reads them creatively, even disruptively, so that they acquire a "popular" resonance (Fiske 1989, 32–43). By "popular" Fiske means "resistant to hegemony," and while I have my reservations about this shopworn typology, the idea of producerly behavior is a relevant heuristic for understanding how texts can generate multiple responses once they reach the brains of creative individuals. This was clearly evident in the pro-Sadler and anti-Sadler responses to "Ballad of the Green Berets."

From the antiwar camp came two protest classics, Pete Seeger's "Waist Deep in the Big Muddy" and Country Joe and the Fish's "I Feel Like I'm Fixing to Die Rag"—the latter, by the way, just as popular among troops in Vietnam as among hometown protesters. These were galvanizing vehicles for the antiwar movement, but neither of them explicitly mentioned Sadler. To my knowledge, the first song to do that was an antiwar crowd-pleaser by the Cambridge-based folksinger Jaime Brockett. It was called "Talking Green Beret New Super Yellow Hydraulic Banana Teeny Bopper Blues." The bloated title was merely Brockett goofing on pop motifs; the song had nothing to say about bananas or teeny boppers or, for that matter, the blues. It did have something to say about Green Berets.

Dedicated to listeners of AM radio, the song develops a clever contrast between music that's "easy to dance to"—a stock phrase on Dick Clark's *American Bandstand*—and music that "people want to march to." Assessing the danceability of the "Ballad" at "about a 73"—a dismal showing in *Bandstand* terms—Brockett imagines the show's teenagers not dancing to it but marching while carrying flags, eating apple pie, and "trading pictures of their moms." When arch hawk Spiro Agnew condemns the "unwashed hippie Marxists" for failing to appreciate the "Ballad," Brockett rates the vice president's IQ also at "about a 73." Then, after failing to sell his own songs to a march-infatuated record producer, he inexplicably joins the Green Berets. His drill sergeant, a "fat little man with a Girl Scout beanie," issues him a twelve-string rifle and tells him, "Kid, I'm gonna teach you how to kill" (Brockett 2005 [1971]).

From the other side of the barricades came songs lauding the Green Berets. Two of them—Lesley Miller's "He Wore the Green Beret" and Craig Arthur's "Son of a Green Beret"—echoed Sadler's notion that self-sacrificing fathers should have self-sacrificing sons. A third song, "When the Green Berets Come Home," was written by a Special Forces captain, Ty Herrington, who later lost his life outside Plie Khan (*Next Stop Is Vietnam* 2011). Their appearance demonstrates that Sadler's "Ballad" was capable of eliciting positive feedback. But the cultural impact of that feedback was not very noticeable, as none of these three songs went anywhere commercially. And, despite their philosophical debt to the singing sergeant, none of them mentioned Sadler's song by name or echoed either his lyrics or his tune.

The first song to do that—I mean to do it with a positive spin—was a 1966 single by the Detroit band Doug and the Omens, featuring an aspiring front man named Bob Seger. In 1968 Seger would question the war in a song called "2 + 2," but in 1966, not yet soured on US policy, he entered the national debate with a parody called "The Ballad of the Yellow Beret." Set to Sadler's music and introduced as "a protest against protesters," this broad slap at the antiwar movement painted draft dodgers as "fearless cowards of the USA" and as "men who faint at the sight of blood" whose "high-heeled boots weren't meant for mud." Its last stanza turned the original Green Beret's dying request into a plea for cross-generational cowardice:

Put a yellow streak down my son's back
Make sure that he never ever fights back
At his physical have him say he's gay
Have him win the yellow beret

Even if one overlooks the homophobic sniggering here, "The Ballad of the Yellow Beret" was a simplistic bagatelle, displaying not so much a pro-war as an anti-antiwar sentiment. While that clearly placed it in the same political camp as Sadler's ballad—it was more homage than attack—Sadler himself didn't see it that way. He issued a cease and desist order, and the song was withdrawn shortly after it was released.

POACHING IN COUNTRY

Perhaps because of the cease and desist order, it would be decades before another American recording artist put new words to Sadler's melody. But if music executives were wary of potential litigation, the same could not be said of young men in uniform. In Vietnam, as in previous war zones, GIs created a rich repertoire of occupational folksongs—noted as early as 1966 by *New York Times* reporter Joseph Treaster. Nearly 200 of them were recorded by US general Edward G. Lansdale and deposited in the Library of Congress's folksong archive (1967, 1967–72). As studies by Les Cleveland (1985, 2003), Lydia Fish (1989, 1993), and Joseph Tuso (1990) show, many of them were "producerly" adaptations of traditional and popular tunes.

For example, Toby Hughes's "Tchepone," which Fish calls "the most popular song of the air war," borrowed the tune of the cowboy classic "Strawberry Roan" (*In Country*, 1991; cf. Guilmartin 1998). Bull Durham's "Danang Lullaby" and the Cosmo Tabernacle Choir's "Montagnard Sergeant" followed the melody of "My Bonnie Lies over the Ocean." And several songs—notably Dick Jonas's "Ubon Tower" (itself a variant of a Korean War song)—were sung to the tune of "Wabash Cannonball." Popular music, too, provided inspiration, as in Jonas's fighter pilot version of Petula Clark's "Downtown" and the creative layering performed by his fellow airman Chip Dockery on such contemporary hits as "Michael, Row the Boat Ashore," "King of the Road," and "Dock of the Bay" (*In Country* 1991). The producerly work that Vietnam servicemen performed on Sadler's hit fit squarely into this creative tradition.

Some of these in-country adaptations were bluntly bellicose. A fighter squadron whose members called themselves the Crusaders, for example, collected their unit songs in a mimeographed booklet, "The Crusader Hymnal." (Horntip n.d.). Their fight song, set to Sadler's music, rejoices in the incineration of an enemy ground force:

> They counted casualties til ten
> The final count was 1000 men

No more they'll pillage, kill, and rape
Cause we fried 'em with the nape

"Nape" was of course napalm—domestically one of the most reviled instruments of the war, but one much admired by imperiled GIs.

This particular parody reflects a black humor that surfaces often in military folksong, as an ironic exculpation, perhaps, by soldiers distressed at participating in brutal enterprises. In Vietnam, Tuso (1990) notes that parodies of Bob Dylan's "Blowing in the Wind" and of "The Streets of Laredo" revealed the ambivalence of pilots flying defoliation missions, while a parody of the 1955 pop hit "Wake the Town and Tell the People" began with this chilling verse:

Strafe the town and kill the people
Drop your napalm in the square:
Do it early Sunday morning.
Catch them while they're still at prayer.

In a similarly macabre vein, the imagined loss of a Vietnamese girl in a Vietcong attack inspired a Sadler parody called "The Ballad of the Green Brassiere" (Tuso 1990, 38):

Now let me tell you about this girl,
She's a true Vietnam pearl
She wore a flower above her ear
And on her chest the Green Brassiere

A VC shell came from above,
Only one thing to remind us of
This little girl we loved so dear
A slightly tattered Green Brassiere

Put silver wings upon her stone
To let her know she's not alone
We love the maid who's buried here
The girl who wore the Green Brassiere

Less grimly, Sadler's tune also inspired a celebration of Air Force planes called "The Inventory." Its verses recorded the virtues (and limitations) of individual birds—the Jolly Green, the Super Spud, the B-52, and so on—while the chorus applauded the valor of their crews:

Silver wings that are no more,
Camouflaged because of war.
Men will die, but don't forget,
They're all a part of our freedom threat! (Tuso 1990, 110–11)

Not all parodies referenced the front-line fighting men. The Twelfth Tactical Fighter Wing's songbook, for example, included Myke Mather's whimsical homage to a public information officer, or PIO—one of those intrepid functionaries whom Sadler himself called "much harassed men who have to deal with all manner of odd problems, with little chance of keeping everybody happy" (Sadler 1967, 147–48). Their marginalization was evident in Mather's lyrics (Horntip n.d.):

There he goes, the PIO
Last to know, first to go
100 times he flies the Hueys
Flown by publicity seeking Lueys

Out to battle he must go
Sent by those in the know
He may take a sniper's round
And be left upon the ground

Fighting men may pass him by
And when they ask, Who was that guy?
I dunno, it's hard to say.
What the hell, Just let him lay.

And when he gets to the golden gate
St. Peter says, You've goofed up, mate!
So go to Hell in all your glory,
When you get back, you can do your story.

A more humorous appropriation, "The Ballad of the Green Grammarians" (Vietnam Veterans Oral History and Folklore Project), was written by Foreign Service officer Dolf Droge. It praised those who learned Vietnamese at the government's Foreign Service Institute and went to Vietnam "trained to fight linguistically":

Their accents clear, their tones are true
Instructors here have taught us what to do

Communicate interminably
And so you'll win over the old VC

Another humorous example, "The Ballad of the ASA," was written in 1966 by members of the Army Security Agency. Like PIOs, soldiers assigned to the ASA might be seen as in the war but not of it—not as brazenly in harm's way as fliers or infantrymen. Their adaptation of Sadler's song showed how sensitive they were to that battlefield fiction:

Drunken soldiers, always high
Dropouts from old Sigma Phi
Men who bullshit all the way,
These are the men from the ASA

Plastic cans upon our ears,
We've been cleared and we're not Queers
One Hundred Men we'll test today,
But only three make the ASA

Trained to go from bar to bar,
That's the life that's best by far
Men who drink will seldom fight,
And the ASA drinks through the night

Black is for the night we fear,
Blue the water we don't go near.
White is for the flag we fly,
Yellow is the reason why.

Red is for the blood we've shed.
As you see, there is no red!
One hundred men re-upped today.
Not a one for the A-S-A!

This is ingenious self-parody—a comment not so much on Green Berets as on the poor reputation that ASA men endured through much of that organization's history. Founded in 1945 as an intelligence arm of the cold war, the Army Security Agency recruited the service's most intelligent soldiers (as measured by IQ tests), gave them top-secret clearance, and made them linguists, cryptologists, and intelligence analysts. In Vietnam, many of them worked directly with Special Forces units, and—despite what the lyrics suggest—they were not a protected species: one of the war's first combat casualties was ASA specialist James Davis. What we hear in these lyrics, then, is what I have elsewhere called "parodic parry" (Tuleja 1997, 9), where a stigmatized group, by embracing the stigma, turns it against an attacker with humorous effect.

In a different kind of parrying, some in-country parodies responded to the notion of Green Beret exceptionalism by trumpeting the valor of rival military units. Here are three examples from the archives of the Vietnam Veterans Oral History and Folklore Project. The first, Jim Hatch's "Mag-16," praised a Marine Aircraft Group:

We are the men of Mag-16
Dirty rough and fighting mean
From way down south we came this way
Don't give a shit about the Green Berets

We fly on missions and fly-aways
Medevacs both night and day
One hundred men were hauled today
Not a goddamned one had a green beret

We too have wings upon our chests
And can compete for America's best
Our wings are gold, we have caps of green
Rather than a Beret I'll be a Marine

In a similar tone, "EOD's Answer to the Green Berets" celebrated the bravery of the Explosive Ordnance Disposal unit, charged with removing explosives from Saigon hotels:

We sing our praises loud and clear
Cause we don't have no PIO near
We'll drink our beer and draw our pay
But we don't wear a darn beret

Tell St. Peter at the golden gate
You're EOD but you're too late
We're all filled up, we're in a daze
The place is full of darn berets

Our tale is done our story told
We're neither brave nor very bold

We'll do our duty come what may
And never wear a darn beret

A final example, written by US Army captain Hershel Gober, twits the bravado of Sadler's comrades by humorously proposing discretion as the better part of valor:

Frightened soldiers from the sky
Screaming hell, I don't wanna die
You can have my jump pay
I'm a little chicken any old way

Silver wings upon their chest
These are men, America's best?
One hundred men will jump today
And only three wear the green beret
(The rest will have parachutes)

Finally, the most sophisticated example of the "We're brave, too" genre—the adaptation called "Green Flight Pay" (sometimes known as "Silver Wings"). It was performed in Vietnam by the GI singing group the Merrymen (cited in Fish 1993).

Silver wings upon my chest,
I fly my chopper above the best.
I can make more dough that way,
But I don't need no green beret.

Tennis shoes upon his feet,
Some folks call him "Sneaky Pete."
He sneaks around the woods all day,
And wears that funny green beret.

It's no jungle floor for me,
I've never seen a rubber tree.
A thousand men will take some test,
While I fly home and take a rest.

And while I fly my chopper home,
I leave him out there all alone.
That is where Green Berets belong,
Out in the jungle writing songs.

And when my little boy is grown,
Don't leave him out there all alone.
Just let him fly and give him pay,
'Cause he can't spend no green beret.

And when my little boy is old,
His silver wings all lined with gold,
He'll also wear a green beret,
In the big parade St. Patrick's Day!

Here humor meets a grim acceptance of war's reality, and it appears in a familiar context: the ongoing competition between ground and air forces that has been a part of military folklore since the invention of the airplane. Here the rivals are the same ones that animate such in-country classics as John Myers's "Green Beret and Friendly FAC" (*In Country* 1991), but with the roles of hero and feckless ally reversed. Since the Merrymen were helicopter pilots, the butt of the song is the plodding, underpaid infantryman—one who, despite his green headgear, must still live like a common foot soldier. Cynicism undercuts the sacrificial theme of the "Ballad," suggesting that the jungle crusade is ultimately about money and hinting at the warrior's fear that he will be left in enemy territory "out there all alone."

The bitterness of this observation is softened by satiric jabs at the Green Berets—men who "take some test" and live in the jungle "writing songs"—and by the final joke about St. Patrick's Day; but this only partially veils the emotional thrust of the song, which seems to me a bouquet-in-disguise for respected comrades. Of all the in-country adaptations I've seen, "Green Flight Pay" strikes me as the most richly intertextual and honestly—if humorously—reflective. It poses as a flyer's dismissal of Special Forces, and there's no question that it reflects what Eric Eliason reminds me is the fierce competition for status between these two elites. At the same time the song's imagery makes it clear that even airmen—notoriously the least humble of fighting men—are able to honor the commitment of their brothers on the ground.

NOT FADE AWAY

Barry Sadler, suffering from a combat wound, was honorably discharged from the service in 1967. He moved with his family to Nashville, where he attempted unsuccessfully to record a second hit and developed a reputation as a mysterious, hard-drinking tough guy—very much the cinematic soldier of fortune. By the early 1970s, he had exhausted his fifteen minutes of fame. He wrote several novels under the series title *Casca the Eternal Mercenary*, about the adventures of the Roman soldier who had pierced the crucified Jesus's side with his lance and was thereafter condemned to endless wandering. These brought him a hefty income, which he promptly squandered. In 1978 he spent a month in prison for the involuntary manslaughter of a romantic rival, and in 1983 he moved to Guatemala, where unsubstantiated rumors had him training insurgents. Five years later, in obscure circumstances, he was shot in the head and airlifted back to the States, where he struggled for a year with brain trauma before dying in October 1989. Before he died he had the pleasure of knowing that his sons, Thor and Brandon, were both intending to join the Special Forces. To that extent, at least, the "Ballad"'s promise was fulfilled. (For Sadler's post-"Ballad" life, see Sipchen 1989.)

But its author's legacy reached beyond the personal. Despite (or perhaps because of) his adulation of sacrifice, his "Ballad" was adopted by Special Forces as an unofficial anthem. Today it remains so meaningful to that elite unit that Green Berets still rise to their feet when it is played, as, for example, at comrades' funeral services. And it is not only US forces who have found it appealing. A YouTube search reveals that at least half a dozen Special Forces units from other nations have adopted the American song and adapted its lyrics. (It has also inspired numerous foreign-language parodies, not all of them respectful of Sadler's sentiments.)

And it has continued to resonate in popular culture. Filmmakers especially have been attracted to it as a patriotic touchstone. Sometimes the touchstone functions simply as a thematic marker: in the Eddy Murphy vehicle *Showtime* (2002), for example, it serves as background music at a gun show. More often, however, the "Ballad" is satirized as shorthand for bravado or jingoistic pride. In 1979's *More American Graffiti,* for example, it accompanies a scene in which a soldier desperate to get out of Vietnam first attempts to shoot himself in the arm and then, when his gun goes off accidentally, comes under withering friendly fire. In the following year's *Caddyshack,* Chevy Chase, as a golf course groundskeeper, sings the "silver wings" stanza under his breath as he drops dynamite into a troublesome gopher's hole. A similar quote occurs in *Jesus's Son* (1999) when an itinerant ne'er-do-well kills a rabbit to the background strains of the "Ballad." In these instances, we are asked simultaneously to laugh at the bathetic antihero and to trivialize the song.

A more politicized appropriation occurred in the 1986 *Saturday Night Live* skit "The Mute Marine." Here William Shatner, as Oliver North—the Marine officer who refused to answer congressional questions about his support for Nicaraguan Contras—stands dumb and proud before a huge flag, as Sadler's melody is given the following lyrics:

Fighting soldier in Vietnam
The perfect son to any mom
He's one part man, one part machine
He's Ollie North, the Mute Marine

Mined the harbor of Managua
Planned the invasion of Grenada
But soon cruel fate would intervene
And he'd become the Mute Marine

He traded arms with Iran
For hostages—what a great plan!
The chances for success were zero
And yet he's still a national hero

He'd like to talk but cannot speak
His will is strong, his case is weak
We may never know just what he's seen
The man they call the Mute Marine

The animus here is directed against North rather than Sadler, yet there's more than an incidental conjunction between the Marine and the Green Beret, and the lyrics themselves imply contempt not just for North's "arms for hostages" deal, but for the Vietnam War experience that preceded it. Here the "Ballad" is associated, as it was in the 1960s, with mindless patriotism.

That association reached a watershed in the 1990s, with two weirdly similar Hollywood productions. In the first, Michael Moore's 1995 spoof *Canadian Bacon*, a US president down in the polls plots to regain his popularity by invading "socialist" Canada. The dim, exuberant sheriff of Niagara Falls, New York, has his neighbors guard such critical sites as a bowling alley and the local bar while the soundtrack plays Sadler singing the "Ballad." Lest we miss the implied critique of Reagan-era posturing, the invasion then begins, spearheaded by "Omega Force" soldiers in red berets.

The "Ballad" appears more obliquely in Barry Levinson's 1997 film *Wag the Dog*. Here, to deflect public attention from a presidential sex scandal, a PR expert and a Hollywood producer invent a US invasion of Albania, produce film footage of it in a studio, and create a crack military force, Unit 303, to serve as the heroes of the fictional operation. "War is show business," claims the flack, and events prove him right, as the American public buys the sham conflict and reelects the president. The film ends with a military funeral for a fallen hero, as the soundtrack plays a song called "God Bless the Men of the 303." Complete with drum rolls and a soaring chorus, this is a clever parody, written by Huey Lewis, of Sadler's song.

The melody is different enough to avoid prosecution, but there's no question that the song, like the movie, takes a less than affectionate look at the Green Berets. As they bury their fallen comrade (he's actually a psychotic criminal who is killed during a rape attempt), the funeral detail of the 303 looks very much like a group of Special Forces soldiers, except that their berets are half green and half leopard skin. Their anthem contains these stanzas:

> We're gathered here with a cross to bear
> The bravest men anywhere
> That this great land will remain free
> God bless the men of the 303
>
> Years from now, when we are gone
> Our children's kids will hear this song
> Think how strong and proud they'll be
> Grandpa fought for the 303

In this song Sadler's devotion to sacrifice has morphed into self-congratulatory blather. Like other 1990s vehicles, "God Bless the Men of the 303" reflected liberal Hollywood's suspicion of Reagan's saber-rattling. But it was also a late skirmish in the ongoing conflict between the lovers and the critics of Barry Sadler.

MANNING UP

The battle continues. Four and a half decades after the "Ballad"'s debut, it remains a touchstone for debate about the virtues of war. Much of this debate is aired on YouTube, which contains numerous videos of "Ballad" performances, among them one of Sadler on national TV. In the three years that this video has been up, it has accumulated a third of a billion views and, at latest count, nearly 10,000 comments.

Some of these are merely personal observations about the poster's association, or lack thereof, to the military, the Vietnam War, or Barry Sadler. "I served in Vietnam and I count many Green Berets among my friends." "I never served in the military but I respect those who do." "My uncle met Barry Sadler in Guatemala and said he was a cool guy." Other comments are more partisan, and these generate quarrels of a spectacularly crude and frequently ad hominem nature. "Flaming," of course, is common among YouTube viewers, as are adolescent mano a mano exchanges; but the level of animosity inspired by the Sadler video—on both sides—is still often exceptionally vituperative.

Within this puerile flak storm, however, it's possible to discern the outlines of an ongoing debate—a debate that emerged from, and encapsulated, a national crisis, and that now, as America is again engaged in policing global conflicts, is as likely to reference Afghanistan or Iraq as Vietnam. Whatever the references, the passion for "our boys in arms" remains consistent, as does the animus against "wimps" who would question their service. Strong feelings energize comments on Sadler's song as well as reactions to a YouTube revival of Seger's "Yellow Beret." My nonscientific sampling suggests that self-proclaimed patriots today far outnumber their detractors and that, on YouTube at least, expressing anything vaguely akin to pacifism will get you instantly denounced as a treasonous swine.

Some comments, to be sure, provide "producerly" readings of the "Ballad." "Jingoistic clap trap," one poster calls it. "One of the most homo-erotic songs ever written," taunts another. And, from a poster with the screen name esrapk, "Soldiers are hired killers for the state." But such interventions are rare, and they are routinely met with crudely aggressive rejoinders. "Fuck u," writes leon687 to esrapk. "They are paid liberators they don't kill because they want to they kill to free the oppressed." Or, from Zallen 1, evoking that seemingly unkillable bogeyman, the 1960s hippie, "Do something useful . . . like paint . . . or whatever you hippies do. God bless our fighting men, and God bless America." From Rotmaster we get this rhetorical question: "Why is it on these green beret songs I always find assholes in the coments who don't apriciate the sacrafices made by the soldiers." And from aKaMumbles, an example of the homophobic posturing that is so common a leitmotif in these sandbox battles: "Me and some buddies sang this at our breakfast some cold afternoon in some hipster coffee shop in West Philadelphia. I've never seen such a pretty song scare the hell out of so many nancy-boys. Long live Barry Sadler."

The reference to "nancy-boys" is not incidental. If there is a single organizing motif to many of the pro-"Ballad" comments, it is the notion that being suspicious of the military makes you a homosexual—that is, in macho folk belief, not a "real" man. What is at issue in many of the YouTube exchanges is not merely what it means to be a good American but what it means to be an American male. In this, we are back to Identity 101—and to the real, emotional challenge of Sadler's musical salvo.

Anxiety over male identity had been a subtext of the "Ballad" wars even in the 1960s. In Sadler's lyric, "only three" of 100 candidates were man enough to pass the virility test, and as we've seen bemused outrage at that boast (SF acceptance rates are undeniably selective, but not that selective) had driven such in-country parodies as "Green Flight Pay" and "EOD's Answer," where the implicit message was "We are warriors, too, and we don't need stupid hats to prove it." A half century later, the question about "real" manliness is still alive, appearing with campy venom in two "Ballad" parodies about gays in the ranks. The first one, "Flaming Fairies," emerged in the Don't Ask, Don't Tell (DADT) era It celebrates not the courage of closeted soldiers but the stereotypical mincing of feminized men, "Bill Clinton's pink berets," focused on promiscuity and unbroken nails. Some representative lines (Fish 2011):

Flaming fairies, we are so shy
I broke a nail, oh I could cry
We are so proud to be this way
We are the fairies of the pink beret

I touch you there and everywhere
You touch me back, ooh I don't care
One hundred men we'll do today
By little old gays of the pink beret

The enemy is like an awful brute
But really, dear, I think they're cute
I'll take him on in a foxhole there
Cause my nylons are the wash 'n wear

A more obviously homophobic version of this parody appears on the anti-
Clinton website William the Impeached (http://williamtheimpeached.
com). Here DADT isn't a disingenuous dodge but a conscious strategy to
infiltrate the ranks:

> Bill Klinton's words upon my ears:
> "You guys have rights, be proud you're queers."
> I once was scared, now I'm OK
> 'Cause I am a fag in the Queen Berets

The same taunt, minus the venom, appears in a more recent parody, "Ballad
of the Queen Beret," by the Washington-based comedy troupe the Capitol
Steps. Here DADT is about to run its course, allowing openly gay soldiers
to celebrate their liberation in lines like these:

> Fighting in Afghanistan
> Fearless men both buff and tan
> He could kick your butt today
> We are the men of the Queen Beret
>
> Brave young men now fill our ranks
> They decorate and paint our tanks
> Join the Navy? There's no way
> Can't wear white past Labor Day
>
> From Iraq to Khyber Pass
> You can bet I'll watch your ass
> The enemy now runs away
> Our battle cry's "YMCA"

Mickey Weems argues in chapter 7 of this volume that parodies like these
may actually hasten the acceptance of gay warriors by opening a space for
campy celebration. Others, of course, find such send-ups insensitive. In any
case, the "pink" and "queen" takes on Sadler's ballad indicate clearly that
the battle for male identity is an ongoing struggle, and that the "normative" male is a fragile construct: "Masculinity is a project never complete"
(Mechling 2005, 218; cf. Kimmel 2001).

I said at the outset that, from its debut, "The Ballad of the Green Berets"
served as a litmus test for people's thinking about the war. But thinking
about the war, any war, is always also a way of thinking about manhood.
This was as important to the meaning of the song as the motif of sacrifice.

It is young men, by and large, who are called upon to sacrifice themselves in war; and their willingness to do so fearlessly is, in our culture as in many others, central to the establishment of their gendered identities. Vietnam-era peaceniks knew this as well as anyone, and their (our) uneasiness with that knowledge may, I suggest, help to explain disdain for Barry Sadler. Meanwhile, on the other side of the barricades, many young men who had never held an M16 invested in Sadler and his brothers in arms—including the cartoon Rambo—every fantasy and fear about their own virility. Then, as now, one way of advertising one's worth as a male was to claim vicarious comradeship with a warrior elite.

In his memoir of America's second incursion into Iraq, National Guardsman John Crawford writes that, for years before he was deployed, "I continued to wonder, as all men do, how I would deal with the bear of war" (2005, xii). It's a telling observation. For young men, whatever their attitudes toward or experience with combat, the hidden question is always "Am I man enough?" As the YouTube debates about Barry Sadler's ostensibly dated song make clear, that question is as relevant today as it was forty years ago. Until it becomes irrelevant, we may expect his "Ballad" to survive.

ACKNOWLEDGMENTS

I am grateful to Lydia Fish, director of the Vietnam Veterans Oral History and Folklore Project, for pointing me to several of the parodies discussed in this chapter. My thanks are due also to her colleague Gary Lee, who at Professor Fish's suggestion made me a CD containing in-country parodies from the project's archives.

WORKS CITED

Brockett, Jaime. 2005 [1971]. *Remember the Wind and the Rain*. EMI/Collectors' Choice Music.

Cleveland, Les. 1985. "Soldier's Songs: The Folklore of the Powerless." *New York Folklore* 11: 79–97.

Cleveland, Les. 2003. "Songs of the Vietnam War: An Occupational Folklore Tradition." *New Directions in Folklore* 7.

Collins, Ace. 2003. *Songs Sung Red, White, and Blue: The Stories behind America's Best-Loved Patriotic Songs*. New York: HarperCollins.

Crawford, John. 2005. *The Last True Story I'll Ever Tell: An Accidental Soldier's Account of the War in Iraq*. New York: Riverhead Books.

Fish, Lydia. 1989. "General Edward G. Lansdale and the Folksongs of Americans in the Vietnam War." *Journal of American Folklore* 102 (406): 390–411. http://dx.doi. org/10.2307/541780.

Fish, Lydia. 1993. "Songs of Americans in the Vietnam War." http://faculty.buffalostate. edu/fishlm/folksongs/americansongs.htm.

Fish, Lydia. 2011. Personal communication.

Fiske, John. 1989. *Understanding Popular Culture*. London: Unwin Hyman.

Guilmartin, John. 1998. "'Tchepone': A Fighter Jock Song." http://faculty.buffalostate. edu/fishlm/folksongs/tchepone.htm.

Halberstadt, Hans. 1988. *Green Berets: Unconventional Warriors*. Novato, CA: Presidio Press.

Horntip, Jack. N.d. "Fighter Pilot and Aviation Songs." In *Folklore from the Jack Horntip Collection*. www.horntip.com.

In Country: Folk Songs of Americans in the Vietnam War. 1991. CD produced by the Vietnam Veterans Oral History and Folklore Project, with liner notes by Lydia Fish. Flying Fish Records.

Kimmel, Michael S. 2001. "Masculinity as Homophobia: Fear, Shame, and Silence in the Construction of Gender Identity." In *Men and Masculinity: A Text Reader*, ed. Theodore F. Cohen, 29–41. Belmont, CA: Wadsworth Thomson Learning.

Lansdale, Edward. 1967. "In the Midst of War." General Edwin Lansdale Songs of the Vietnam War. Library of Congress. American Folklife Center Archive of Folk Culture. AFC 1975/014.

Lansdale, Edward. 1967–1972. General Edwin Lansdale Collection of Vietnam War Songs. Library of Congress. American Folklife Center Archive of Folk Culture. AFS 18,977–18,982.

Mechling, Jay. 2005. "The Folklore of Mother-Raised Boys and Men." In *Manly Traditions: The Folk Roots of American Masculinity*, ed. Simon Bronner, 211–27. Bloomington: Indiana University Press.

Next Stop Is Vietnam: The War on Record 1961–2008. 2011. Bear Family Records CD box set.

Prados, John. 2009. *Vietnam: The History of an Unwinnable War, 1945–1975*. Lawrence: University Press of Kansas.

Sadler, Barry. 1966. *Ballads of the Green Berets*. RCA Records.

Sadler, Barry. 1967. *I'm a Lucky One*. New York: Macmillan.

Sipchen, Bob. 1989. "The Ballad of Barry Sadler." *Los Angeles Times*, January 27, sec. 5: 1–4.

Treaster, Joseph. 1966. "The GI's View of Vietnam." *New York Times Magazine* (October 30): 100ff.

Tuleja, Tad. 1997. "Making Ourselves Up: On the Manipulation of Tradition in Small Groups." Introduction to *Usable Pasts: Traditions and Group Expressions in North America*, ed. Tad Tuleja. Logan: Utah State University Press.

Tuso, Joseph F. 1990. *Singing the Vietnam Blues: Songs of the Air Force in Southeast Asia*. College Station: Texas A&M University Press.

Vietnam Veterans Oral History and Folklore Project (VVOHFP). Archives housed at Buffalo State College and maintained by project director Lydia Fish.

William the Impeached website. http://williamtheimpeached.com. http://www.arrse.co.uk/ wiki/Multi_Letter_Acronyms.

Selected Bibliographies

The following bibliographies are not intended to be exhaustive listings of works on military folklore in general but rather to present those sources most useful and accessible to students of such lore in English-speaking military units since the First World War.

GENERAL AND MISCELLANEOUS

The items in this section address military customs, traditions, or culture in general, or they cover multiple expressive forms (for example, song and slang).

Boatner, Mark Mayo. 1976. *Military Customs and Traditions*. Westport, CT: Greenwood.

Bronner, Simon J. 2006. *Crossing the Line: Violence, Play, and Drama in Naval Equator Traditions*. Amsterdam: Amsterdam University Press. http://dx.doi.org/10.5117/9789053569146

Burke, Carol. 1996. "Military Folklore." In *American Folklore: An Encyclopedia*, ed. Jan Harold Brunvand. New York: Garland.

Burke, Carol, ed. 2003. "Special Issue: Military Folklore." *New Directions in Folklore* 7. With a Special Editor's Introduction by Carol Burke.

Burke, Carol. 2004. *Camp All-American, Hanoi Jane, and the High-and-Tight: Gender, Folklore, and Changing Military Culture*. Boston: Beacon Press.

Cleveland, Les. 1987. "Military Folklore and the Underwood Collection." *New York Folklore* 13 (3–4): 87–103.

Dorson, Richard. 1977 [1959]. *American Folklore*. Chicago: University of Chicago Press. Section on "GI Folklore": 268–276.

Edwards, Thomas Joseph. 1950. *Military Customs*. Aldershot, England: Gale & Polden.

Field, Cyril. 1939. *Old Times Under Arms: A Military Garner*. London: William Hodge and Company.

Fish, Lydia. 2003. "Informal Communications Systems in the Vietnam War: A Case Study in Folklore, Technology, and Popular Culture." *New Directions in Folklore* 7 (Special Issue: Military Folklore).

Keith, Sam. 1950. "The Flying Nightmares." *New York Folklore Quarterly* 6 (3): 154–60.

Koch, Edwin E. 1953. "G.I. Lore: Lore of the Fifteenth Air Force." *New York Folklore Quarterly* 9:59–70.

Lally, Kelly A. 1987. "Living on the Edge: The Folklore of Air Force Pilots in Training." *Midwestern Folklore* 13 (2): 107–20.

Leary, James P. 1975. "Folklore and Photography in a Male Group." In *Saying Cheese: Folklore and Visual Communication*, ed. Steven Ohrn and Michael E. Bell, 13: 45–50. Bloomington, IN: Folklore Forum Bibliographic and Special Series.

Lovette, Leland. 1939. *Naval Traditions and Usage*. 3rd ed. Annapolis: US Naval Institute.

Sandels, Robert. 1983. "The Doughboy: The Formation of a Military Folk." *American Studies* (Lawrence, Kan.) 24 (1): 69–88.

Stevens, Bob. 1975. *There I Was . . . Flat on My Back*. Fallbrook, CA: Aero Publishers.

Thorpe, Peter. 1967. "Buying the Farm: Notes on the Folklore of the Modern Military Aviator." *Northwest Folklore* 2 (1): 11–7.

Underwood, Agnes Nolan. 1947. "Folklore from GI Joe." *NYFQ* 3:285–97.

SPEECH

Bay, Austin. 2007. *"Embrace the Suck": A Pocket Guide to Milspeak*. New York: New Pamphleteer.

Burke, Carol. 2003. "Military Speech." In *New Directions in Folklore*, Special Issue: Military Folklore, edited by Carol Burke 7.

Burns, Richard Allen. 2003. "'This Is My Rifle, This Is My Gun': Gunlore in the Military." *New Directions in Folklore* 7 (Special Issue: Military Folklore).

Clark, Gregory R. 1990. *Words of the Vietnam War*. Jefferson, NC: McFarland.

Colby, Elbridge. 1942. *Army Talk: A Familiar Dictionary of Soldier Speech*. Princeton, NJ: Princeton University Press.

Cornell, George. 1981. "G.I. Slang in Vietnam." *Journal of American Culture* 4 (2): 195–200. http://dx.doi.org/10.1111/j.1542-734X.1981.0402_195.x.

Cragg, Dan. 1980. "A Brief Survey of Some Unofficial Prosigns Used by the United States Armed Forces." *Maledicta: The International Journal of Verbal Aggression* 4 (2): 167–73.

Dickson, Paul. 2004. *War Slang: American Fighting Words and Phrases since the Civil War*. 2nd ed. Washington, DC: Brassey's.

Elkin, Frederick. 1946. "The Soldier's Language." *American Journal of Sociology* 51 (5): 414–22. http://dx.doi.org/10.1086/219852.

Elting, John, Dan Cragg, and Ernest Deal, eds. 1984. *A Dictionary of Soldier Talk*. New York: Scribner's.

Fraser, Edward, and John Gibbons. 1968 [1925]. *Soldier and Sailor Words and Phrases*. Detroit: Gale Research.

Kenagy, S. G. 1978. "Sexual Symbolism in the Language of the Air Force Pilot: A Psychoanalytic Approach to Folk Speech." *Western Folklore* 37 (2): 89–101. http://dx.doi.org/10.2307/1499316.

Reinberg, Linda. 1991. *In the Field: The Language of the Vietnam War*. New York: Facts on File.

Riordan, John Lancaster. 1946. "American Naval 'Slanguage' in the Pacific in 1945." *California Folklore Quarterly* 5 (4): 375–90. http://dx.doi.org/10.2307/1495930.

Rives, Timothy D. 2003. "The Work of Soldier Poetry in Kansas, 1917–1919." *New Directions in Folklore* 7 (Special Issue: Military Folklore).

Robson, Martin. 2008. *Not Enough Room to Swing a Cat: Naval Slang and Its Everyday Use*. Annapolis: Naval Institute Press. Well annotated, not just definitions.

Roulier, Joseph B. 1948. "Service Lore: Army Vocabulary." *New York Folklore Quarterly* 4:15–28.

Tuleja, Tad. 2011. "'Let God Sort 'em Out': The Archaeology of a Warrior's Catchphrase." Paper presented to the American Folklore Society annual meeting.

NARRATIVES

The items below address various forms of narrative, including popular legends and beliefs, tall tales, and personal recollections of military experience. Additional material of this type may, of course, be found in the memoirs and war stories written by veterans.

Baky, John. 1994. "White Cong and Black Clap: The Ambient Truth of Vietnam War Legendry." In *Nobody Gets Off the Bus: The Vietnam Generation Big Book*, ed.Dan Duffy and Kali Tal. Tucson: Burning Cities Press.

Budra, Paul, and Michael Zeitlin, eds. 2004. *Soldier Talk: The Vietnam War and Oral Narrative*. Bloomington: Indiana University Press.

Burke, Carol. 2006. "Soldiers Real and Imagined and the Stories They Tell." *Contemporary Legend* 4: pages.

Feola, Christopher J. 1989. "The American Who Fought on the Other Side." *New York Folklore* 15 (1–2): 119–120.

Jackson, Bruce. 1990. "The Perfect Informant." *Journal of American Folklore* 103 (410): 400–16. http://dx.doi.org/10.2307/541608.

Jansen, William Hugh. 1948. "The Klesh Maker." *Hoosier Folklore* 7: 47–50.

Lembke, Jerry. 1998. *The Spitting Image: Myth, Memory, and the Legacy of Vietnam*. New York: NYU Press.

Pearson, Barry. N.d. "The Soldier's Point of View: The Experience of World War II and Vietnam as Portrayed in Folklore and Oral History." Unpublished manuscript in "Articles and Papers." http://faculty.buffalostate.edu/fishlm/articles/index.htm.

Pratt, John Clark, ed. 1984. *Vietnam Voices: Perspectives on the War Years, 1941–1982*. New York: Viking.

Rich, George W., and David F. Jacobs. 1973. "Saltpeter: A Folkloric Adjustment to Acculturation Stress." *Western Folklore* 32 (3): 164–79. http://dx.doi.org/10.2307/1498382.

Shorrocks, Graham. 1990. "Body Bag Backlog: A Contemporary Legend?" *Foaftale News* 20:5.

Tuleja, Tad. 2007. "Spit and Spin: Rival 'Memory Narratives' of Veteran Abuse." Paper presented to the American Folklore Society annual meeting.

Wallrich, William. 1960. "Superstition and the Air Force." *Western Folklore* 19 (1): 11–6. http://dx.doi.org/10.2307/1498001.

Wilkinson, Stephen. 1998. "Aviation Legends." Air and Space. Online supplement, December–January 1998 [*sic*].

Yates, Norris. 1949. "Some 'Whoppers' from the Armed Services." Tall tales. *Journal of American Folklore* 62 (244): 173–80. http://dx.doi.org/10.2307/536313.

SONGS

This bibliography focuses on original songs or adaptations of popular tunes created by the troops themselves, although some items cover both such occupational folksongs and the armed forces' favorite popular tunes. It excludes collections of patriotic and march music, government songbooks, and general folksong collections that may contain troop favorites.

Arthur, Max. 2001. *When This Bloody War Is Over: Soldiers' Songs of the First World War.* London: Piatkus Books.

Brophy, John, and Eric Partridge. 1965. *The Long Trail: What the British Soldier Sang and Said in the Great War of 1914–1918.* London: Andre Deutsch.

Burke, Carol. 1992. "'If You're Nervous in the Service . . .': Training Songs of Female Soldiers in the '40s." In *Visions of War: World War II in Popular Literature and Culture,* edited by M. Paul Holsinger and Mary Anne Schofield, 127–137. Bowling Green, OH: Bowling Green State University Popular Press.

Cary, Melbert B., Jr. 1935. "Mademoiselle from Armentieres." *Journal of American Folklore* 47 (186): 369–76. http://dx.doi.org/10.2307/535692.

Cleveland, Les. 1984. "When They Send the Last Yank Home: Wartime Images of Popular Culture." *Journal of Popular Culture* 18 (3): 31–6. http://dx.doi.org/10.1111/j.0022-3840.1984.1803_31.x.

Cleveland, Les. 1985. "Soldier's Songs: The Folklore of the Powerless." *New York Folklore* 11:79–97.

Cleveland, Les. 1994a. *Dark Laughter: War in Song and Popular Culture.* Westport, CT: Praeger.

Cleveland, Les. 1994b. "Singing Warriors: Popular Songs in Wartime." *Journal of Popular Culture* 28 (3): 155–75. http://dx.doi.org/10.1111/j.0022-3840.1994.2803_155.x.

Cleveland, Les. 2003 [1988]. "Songs of the Vietnam War: An Occupational Folk Tradition." In *New Directions in Folklore* 7 (Special Issue: Military Folklore).

Denisoff, Serge. 1973. *Songs of Protest, War, and Peace.* Santa Barbara, CA: ABC-Clio.

Dolph, Edward Arthur. 1929. *Sound Off! Soldier Songs from the Yankee Doodle to Parley voo.* New York: Cosmopolitan.

Dolph, Edward Arthur. 1942. *Sound Off! Soldier Songs from the Revolution to World War II.* New York: Farrar & Rinehart.

Fish, Lydia. 1989. "General Edward G. Lansdale and the Folksongs of Americans in the Vietnam War." *Journal of American Folklore* 102 (406): 390–411. http://dx.doi.org/10.2307/541780.

Fish, Lydia. 1991. Liner Notes to *In Country: Folk Songs of Americans in the Vietnam War.* Chicago: Flying Fish Records.

Fish, Lydia. 1993. "Songs of Americans in the Vietnam War." In "Articles and Papers." http://faculty.buffalostate.edu/fishlm/articles/index.htm.

Fish, Lydia. 1996. "Songs of the Air Force in the Vietnam War." In "Articles and Papers." http://faculty.buffalostate.edu/fishlm/articles/index.htm.

Fish, Lydia. 2003. "Informal Communications Systems in the Vietnam War: A Case Study in Folklore, Technology, and Popular Culture." In *New Directions in Folklore* 7 (Military Folklore special issue).

Getz, C. W. 1981. *The Wild Blue Yonder: Songs of the Air Force.* vol. 1. Burlingame, CA: Redwood Press.

Getz, C.W. 2004. "Rhythm and Blue." *Air Force Magazine* (July): 78–80. Also in "Articles and Papers," http://faculty.buffalostate.edu/fishlm/articles/index.htm.

Guilmartin, John. 1998. "'Tchepone': A Fighter Jock Song." In "Articles and Papers." http://faculty.buffalostate.edu/fishlm/articles/index.htm.

Hamilton, Hamish. 1945. *Ballads of World War II.* Glasgow: Lili Marlene Club.

Hench, Atcheson. 1921. "Communal Composition in the AEF." *Journal of American Folklore* 34 (134): 386–9. http://dx.doi.org/10.2307/534926.

Heuer, Martin. 2003. "Personal Reflections on the Songs of Army Aviators in the Vietnam War." In *New Directions in Folklore 7*, special issue on military folklore, ed. Carol Burke.

Nettleingham, Frederick Thomas. 1917. *Tommy's Tunes.* London: Erskine MacDonald.

Nettleingham, Frederick Thomas. 1919. *More Tommy's Tunes.* London: Erskine MacDonald. British WWI.

Next Stop Is Vietnam: The War on Record 1961–2008. 2011. Bear Family Records CD box set.

Niles, John Jacob, Douglas S. Moore, and A. A. Wallgreen. 1927. *Singing Soldiers.* New York: Charles Scribner's Sons.

Niles, John Jacob, Douglas S. Moore, and A. A. Wallgreen. 1929. *The Songs My Mother Never Taught Me.* New York: Macaulay.

Page, Martin. 1973. *Kiss Me Goodnight, Sergeant Major.* London: Hart Davis MacGibbon.

Page, Martin. 1976. *For Gawdsake Don't Take Me.* London: Hart Davis MacGibbon.

Posselt, Eric. 1943. *Give Out! Songs of, by and for the Men in Service.* New York: Arrowhead Press.

Pound, Louise. 1923. "Hinkie Dinkie Parlevous." *Journal of American Folklore* 36 (140): 202–3. http://dx.doi.org/10.2307/535220.

Tuso, Joseph F. 1990. *Singing the Vietnam Blues: Songs of the Air Force in Southeast Asia.* College Station: Texas A&M University Press.

Vennum, Thomas, Jr., and Mickey Hart, producers. 1997. *American Warriors: Songs for Indian Veterans.* A Mickey Hart Collection CD, Smithsonian Folkways.

Wallrich, William. 1954. "United States Air Force Parodies Based upon 'The Dying Hobo.'" *Western Folklore* 13 (4): 236. http://dx.doi.org/10.2307/1496436.

Wallrich, William. 1957. *Air Force Airs.* New York: Duell, Sloan, and Pearce.

Ward-Jackson, C.H. 1945. *Airman's Song Book.* London: Blackwood.

Ye AEF Hymnal. 1918. Nancy: Berger-Levrault.

York, Dorothea, ed. 1931. *Mud and Stars: An Anthology of World War Songs and Poetry.* New York: Holt.

CADENCES

Burke, Carol. 1989. "Marching to Vietnam." *Journal of American Folklore* 102 (406): 424–41. http://dx.doi.org/10.2307/541782.

Burns, Richard Allen. 2006. "'I Got My Duffel Bag Packed/And I'm Goin' to Iraq': Marching Chants in the Military." In *Ballad Mediations: Folksongs Recovered, Represented, and Reimagined*, edited by Roger deV. Renwick and Sigrid Rieuwerts. Trier: WVT.

Carey, George. 1965. "A Collection of Airborne Cadence Chants." *Journal of American Folklore* 178 (307): 52–61. http://dx.doi.org/10.2307/538103.

Jackson, Bruce. 1967. "What Happened to Jody?" *Journal of American Folklore* 80 (318): 387–96. http://dx.doi.org/10.2307/537417.

Johnson, Sandee Shaffer. 1983. *Cadences: The Jody Call Book, No. 1*. Canton, OH: Daring Press.

Johnson, Sandee Shaffer. 1986. *Cadences: The Jody Call Book, No. 2*. Canton, OH: Daring Press.

Knight, Jeff Parker. 1990. "Literature as Equipment for Killing: Performance as Rhetoric in Military Training Groups." *Text and Performance Quarterly* 10 (2): 157–68. http://dx.doi.org/10.1080/10462939009365965.

Miles, Donna. 1995. "Jodies: Songs on the Move." *Soldiers Magazine* 50 (6): n.p.

Trnka, Susanna. 1995. "Living a Life of Sex and Danger: Women, Warfare and Sex in Military Folk Rhymes." *Western Folklore* 54: 232–241.

MATERIAL CULTURE

Chittenden, Varick A. 1989. "'These Aren't Just My Scenes': Shared Memories in a Vietnam Veteran's Art." *Journal of American Folklore* 102 (406): 412–23. http://dx.doi.org/10.2307/541781.

Dewhurst, Kurt C. 1988. "Pleiku Jackets, Tour Jackets, and Working Jackets: 'The Letter Sweaters of War.'" *Journal of American Folklore* 101 (399): 48–52. http://dx.doi.org/10.2307/540249.

Pershing, Linda, and Nishelle Y. Bellinger. 2010. "From Sorrow to Activism: A Father's Memorial to His Son Alexander Arredondo, Killed in the US Occupation of Iraq." *Journal of American Folklore* 123 (488): 179–217. http://dx.doi.org/10.1353/jaf.0.0136.

Ryan, Bernadene. 2011. "Challenge Coins: Agents of Identity in Negotiating Inclusion into Military Communitas." Poster presentation at American Folklore Society annual meeting.

Santino, Jack. 1992. "Yellow Ribbons and Seasonal Flags: The Folk Assemblage of War." *Journal of American Folklore* 105 (415): 19–33. http://dx.doi.org/10.2307/541997.

Smyth, Cecil. 1988. "Unofficial Military Insignia of the Vietnam War: United States Army Special Forces." *Antiques and Collecting Hobbies* 92:28–30.

Sossaman, Stephen. 1989. "More on Pleiku Jackets in Vietnam." *Journal of American Folklore* 102 (403): 76. http://dx.doi.org/10.2307/540083.

Tuleja, Tad. 1997. "Closing the Circle: Yellow Ribbons and the Redemption of the Past." In *Usable Pasts: Traditions and Group Expressions in North America*, edited by Tad Tuleja. Logan: Utah State University Press.

HUMOR

Cerf, Bennett, ed. 1943. *The Pocket Book of War Humor*. New York: Pocket Books.

Foster, Ted. 1980. *The Vietnam Funny Book*. Novato, CA: Presidio Press.

Melvin, Ken. 1966. *Sorry 'bout That!* Tokyo: Wayward Press.

Myers, James E., ed. 1990. *A Treasury of Military Humor*. Springfield, IL: Lincoln-Herndon.

Zidek, Tony. 1965. *Choi-oi: The Lighter Side of Vietnam*. Tokyo: Charles Tuttle.

Contributors

Eric A. Eliason is a professor of folklore at Brigham Young University. An avid field researcher, he has published on Mormon, Caribbean, Russian, English, Afghan, American, Mexican, and biblical cultural traditions. His books include *The J. Golden Kimball Stories, Wild Games: Hunting and Fishing Traditions in North America* (with Dennis Cutchins), *Black Velvet Art*, and the forthcoming *Latter-day Lore: A Handbook of Mormon Folklore Studies* (with Tom Mould). He is writing the folklore chapter for the forthcoming *Oxford Companion to Mormonism*. From 2002 to 2008 Eliason was the chaplain for the First Battalion, Nineteenth Special Forces in the Utah Army National Guard. He served in Afghanistan, in the Philippines, and at Arlington National Cemetery. He lives in Springville, Utah, with his wife and four children.

Tad Tuleja teaches writing at American University and has published widely in the humanities and social sciences. His books on American subjects include *American History in 100 Nutshells, The New York Public Library Book of Popular Americana*, and the edited volume *Usable Pasts: Traditions and Group Expressions in North America*. The son of a naval officer and military historian, he has a long-standing interest in the anthropology of warfare and has written papers on yellow ribbons, the unofficial slogans of military elites, and the reported abuse of returning Vietnam War veterans. At Harvard, Princeton, Willamette University, and the University of Oklahoma, Tuleja taught courses that examined war and gender, just-war theory, and military memoirs. He holds master's degrees from Cornell and the University of Sussex (England) and a PhD in folklore and anthropology from the University of Texas at Austin.

Carol Burke, professor of English at the University of California, Irvine, holds a PhD in English from the University of Maryland. She combines her ethnographic skills as a folklorist with an interest in literary journalism. Her publications include *Camp All-American, Hanoi Jane, and the High-and-Tight*, a study of military culture; *Women's Visions*, which explores accounts of the supernatural exchanged by women in prison; *The Creative Process* (coauthored with Molly Tinsley), a creative writing text; the collections of family folklore *Plain Talk* and *Back in Those Days* (the latter coauthored with Martin Light); *Close Quarters*, a collection of poems; and articles in the *Nation* and the *New Republic*. She also wrote the "Military Folklore" entry for *American Folklore: An Encyclopedia*, edited by Jan Brunvand; and guest edited the *New Directions in Folklore* special issue on military folklore. In 2008–9, on leave from the University of California, she did fieldwork with a US infantry unit in Iraq; in 2010–11, she spent nine months with US troops in Afghanistan.

RICHARD ALLEN BURNS spent four years in the US Marine Corps after high school, earned two undergraduate degrees, taught high school for four years, then received his PhD in anthropology/folklore from the University of Texas at Austin. He is currently an associate professor of anthropology and folklore at Arkansas State University, where he teaches anthropology and courses that constitute ASU's minor in folklore as well as graduate seminars in folklore that are part of ASU's doctoral program in heritage studies. He is currently pursuing an investigation of folklife in the Arkansas Delta in addition to his ongoing research on military and prison folklore.

ANGUS KRESS GILLESPIE is a folklorist with a strong interest in the traditions of people who live by the sea. He grew up in Patuxent River, Maryland, on the Chesapeake Bay, and graduated from nearby Leonard Hall Naval School with the rank of lieutenant (JG). At college, he honed his skills in navigation and seamanship as a member of the Yale Corinthian Yacht Club, the oldest collegiate sailing club in the world. After graduating from Yale he worked as a technical editor for the Electric Boat Division of General Dynamics before going on to earn a doctorate in American studies from the University of Pennsylvania. He is a member of the Navy League and a professor at Rutgers University, where he offers courses in maritime history and culture. In these courses, students interpret the past not only through readings and classroom discussions but also through studies of maritime artifacts and folklore. He also deals with maritime policy, touching on economic and environmental issues as well as current policy regarding world trade and regulatory reform, conservation and fisheries, national defense, and admiralty law.

LISA GILMAN is associate professor of English and director of the folklore program at the University of Oregon. She has a PhD in folklore with a minor in African studies from Indiana University. Her research explores intersections between expressive forms, gender, and politics. Since 2005, she has been involved in two research projects with young American veterans of the current conflicts in Iraq and Afghanistan. The first explores troops' musical listening practices during deployments; she is currently developing a book proposal on this research. The second has resulted in a documentary film, *Coffee Strong*, about veterans who run an antiwar GI coffeehouse outside of Fort Lewis Army Base in Washington State. These veterans are also the subject of her chapter in this volume.

GREG KELLEY holds a PhD in folklore from Indiana University. Former president of the Hoosier Folklore Society and editor of the journals *Folklore Forum* and *Midwestern Folklore,* he also served as coordinating publisher for Trickster Press, where he was associate editor of Dell Hymes's *Reading Takelma Texts,* Warren Roberts's *Log Cabins of Southern Indiana,* and *Fields of Folklore: Essays in Honor of Kenneth S. Goldstein* (edited by Roger D. Abrahams). He has published broadly on folklore, folklore and literary relations, and humor. Currently, he teaches in the

School of English and Theatre Studies at the University of Guelph and in media studies at the University of Guelph-Humber.

ELINOR LEVY holds a doctorate from Indiana University, Bloomington, where her dissertation explored the use of collaborative research methodologies in folklore. Her interest in the military began with her work on Vietnam veterans memorials for her master's thesis at California State University, Sacramento. While she has no direct connection to the military, she finds the ways in which community is formed and maintained by military units fascinating and vital for mission success. As an adjunct assistant professor at Raritan Valley Community College and an adjunct instructor at Fairleigh Dickinson University, she has taught courses in military folklore, folklore and food, folklore and the media, the anthropology of death, the anthropology of place and memory, applied anthropology, ethnography, and ritual and festival.

JAY MECHLING is professor emeritus of American studies at the University of California, Davis. He earned his BA in American studies at Stetson University and his graduate degrees in American civilization at the University of Pennsylvania. He directed the American studies program at the University of California from 1978 to 1988, was editor of *Western Folklore* from 1984 through 1988, and was elected president of the California Folklore Society in 1991. A fellow of the American Folklore Society since 1998 and a former chair of the California Council for the Humanities, he has received many awards for his achievements in teaching and curricular development, most recently the 2006 Davis Prize, his campus's highest honor for outstanding teaching and scholarship. Mechling has published over 100 essays and articles in books, journals, and encyclopedias. He was coeditor of *American Wildlife in Symbol and Story* and author of *On My Honor: Boy Scouts and the Making of American Youth*. He is one of the three senior editors for the *Encyclopedia of American Studies* and coauthor of *Embellishing Eden: The Hand-Painted Photograph in Florida*, the catalog of an exhibition he curated with Elizabeth Walker Mechling.

JUSTIN M. OSWALD graduated from Pennsylvania State University with a BA in media studies: international communications and a minor in Italian. Raised in Pennsylvania, he moved in 2004 to Virginia to become a special investigator, a position that entailed frequent conversations with military personnel. Since 2006 he has been a personnel security specialist for the US Air Force, engaging in daily contact with enlisted personnel, civilians, and contractors working for the military. He is also currently enrolled in George Mason University's master of arts in interdisciplinary studies program, with a concentration in folklore. He is particularly interested in the study of supernatural belief systems, with an emphasis on legends and death lore.

MICKEY WEEMS is a former US Marine and the cofounder of the Qualia Festival of Gay Folklife. He holds a PhD from Ohio State University, teaches at Columbus

State Community College, and is chief editor of the forthcoming *Encyclopedia of Gay Folklife*. In his ethnography *The Fierce Tribe: Masculine Identity and Performance in the Circuit*, he draws on religious studies and on his practice of Brazilian Candomblé to reveal the spiritual and parodic elements in the gay "Circuit" culture; a shorter study of the same phenomenon appears in *Manly Traditions: The Folk Roots of American Masculinities*, edited by Simon Bronner. His interdisciplinary and cross-cultural work on spiritual traditions has nurtured his passionate dedication to ecumenical tolerance, to world peace, and to the promotion of masculine ideals that reject the celebration of violence.

KRISTI YOUNG holds a master of library science degree from the University of North Texas and bachelor of arts and master of arts degrees from Brigham Young University, where she works in the library system. Her daughter's experience as an Air Force wife stimulated her interest in the uses of narrative among military spouses and led to the interviews underlying her chapter in this book. Among Young's writings on marriage are the article "Courtship" for the *Encyclopedia of Women's Folklore and Folklife* and the articles "Marriage" and "Showers: Wedding and Baby" for the *Encyclopedia of American Folklife*; she is also coeditor, with Mary Jane Woodger and Thomas Holman, of the anthology *Latter-day Saint Courtship Patterns*. Her chapter "Oral History among the Orchards" will appear in the forthcoming anthology *Oral History, Community, and Work in the American West*.

Index